Excel

Get the Results You Want!

Year 9 Mathematics Practice Tests

Allyn Jones

Reprinted 2021, 2023, 2025

ISBN 978 1 74125 636 9

Pascal Press
PO Box 250
Glebe NSW 2037
www.pascalpress.com.au

Publisher: Vivienne Joannou
Project editor: Rosemary Peers
Edited by Rosemary Peers
Answers checked by Peter Little
Typeset by Kim Webber
Printed by Vivar Printing/Green Giant Press

Contents

MEASUREMENT AND GEOMETRY

Pythagoras' Theorem

Area, Surface Area and Volume

Trigonometry

Similarity

Measurement and Geometry Strand Tests

STATISTICS AND PROBABILITY

Statistics

Probability

Statistics and Probability Strand Tests

YEARLY EXAMINATIONS

Introduction

- The main purpose of this book is to provide students with **extensive test and exam practice** at all levels in Year 9, in the topics they cover in class. This way students will be given the test and exam practice they require to be fully prepared to excel at their class tests and exams.
- This book covers all the topics in the **Year 9 Australian Curriculum (Mathematics).**

The following features will help students excel in their class tests and exams:

- This book is a **workbook** that students can write in, ensuring that they have all their tests, exams and working in one place. This is invaluable when revising for topic tests and exams—no lost pages or notes.
- Each unit of work has two **pretests** designed to identify any problems students might have with prior knowledge requirements for the course.
- The **main tests** all have **two sections**: Part A has multiple-choice questions and Part B has short-response questions.
- The main tests also have different **levels of difficulty**.
  - **Level 1** tests are the easiest. They should determine whether a student has a basic understanding of the topic.
  - **Level 2** tests are aimed at the average student. Those who do well in these tests will have a good grasp of the content of the topic.
  - **Level 3** tests are the most challenging. They are aimed at good students and should determine whether they have a thorough knowledge of the content and are able to apply that knowledge.
- This book is therefore useful to **all levels of students,** who can attempt more challenging questions or complete questions that can be answered more easily.
- There are **nine strand tests**: Number and Algebra, Measurement and Geometry, and Statistics and Probability, with multiple-choice and extended-response formats at three different levels.
- There are **three yearly examination papers**.
- All tests and exams are **timed** and a **marking scheme** is provided to give students an idea of their progress.
- **Answers to all questions** are provided at the back of the book.
- Note that the **innovative test and exam format** of this book can be used both as a means of assessing students' knowledge and understanding, and also to provide comprehensive revision of the course.

There are four titles in the series:

Fractions, Decimals, Percentages and Ratios

PRETEST

LEVEL 1

Time allowed: 10 minutes **Total marks: 10**

Marks

1 Which of these is 3.749×100?

(A) 37.49 (B) 374.9 (C) 3749 (D) 374 900 [1]

2 What is the place value of 3 in the number 239.2?

(A) 3 tens (B) 3 units (C) 3 tenths (D) 3 hundredths [1]

3 How many decimal places has the number 712.416?

(A) 3 (B) 4 (C) 5 (D) 6 [1]

4 Which of these is 0.637 21, correct to two decimal places?

(A) 0.6 (B) 0.7 (C) 0.63 (D) 0.64 [1]

5 What is the value of 4.3^2?

(A) 8.6 (B) 4.6 (C) 4.9 (D) 18.49 [1]

6 Liarni used her calculator to evaluate 3.86×9.5. Which of these is the closest to her answer on the calculator?

(A) 36 (B) 37 (C) 38 (D) 39 [1]

7 Which of these is the same as 0.25?

(A) $\frac{2}{5}$ (B) $\frac{1}{25}$ (C) $\frac{1}{4}$ (D) $\frac{5}{2}$ [1]

8 The number below is the display on a calculator. What is the number correct to two decimal places?

8.127395175

(A) 8.12 (B) 8.13 (C) 812 (D) 813 [1]

9 The number one billion is written as 1 followed by zeros. How many zeros are in the number?

(A) 3 (B) 6 (C) 9 (D) 12 [1]

10 The table shows the finishing times for three friends in a race.

Name	Finishing position	Time (seconds)
Grace	fifth	16.8
Mia	fourth	?
Tahlia	second	16.29

Which of these could have been Mia's time?

(A) 16.2 (B) 16.9 (C) 16.25 (D) 16.6 [1]

Total marks

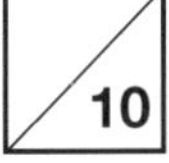

Fractions, Decimals, Percentages and Ratios

PRETEST — LEVELS 2 AND 3

Time allowed: 10 minutes — **Total marks: 10**

Marks

1 Which of these is $\frac{4.271}{8.289}$, correct to two decimal places?

(A) 0.5 (B) 0.51 (C) 0.52 (D) 0.53 — 1

2 What is the missing number?

$6.51 \times \square = 6510$

(A) 10 (B) 10^2 (C) 10^3 (D) 10^4 — 1

3 Increase \$400 by 20%.

(A) \$40 (B) \$420 (C) \$440 (D) \$480 — 1

4 What is the value of 3.86 × 10 000?

(A) 3.860 000 (B) 386 (C) 3860 (D) 38 600 — 1

5 Which of these is the same as $\frac{2}{3}$?

(A) 0.23 (B) 0.6 (C) $0.\dot{2}\dot{3}$ (D) $0.\dot{6}$ — 1

6 Which of these is the same as the ratio of 16 : 100?

(A) 2 : 5 (B) 4 : 25 (C) 5 : 2 (D) 25 : 4 — 1

7 The table shows Mia's results in four tests.

Test	Result
English	15 out of 20
Mathematics	42 out of 50
Science	48 out of 60
History	56 out of 70

In which test did Mia score the best result?

(A) English (B) Mathematics (C) Science (D) History — 1

8 What is the value of $10^3 \times 10^2$?

(A) 10^5 (B) 10^6 (C) 100^5 (D) 100^6 — 1

9 Which of these is arranged in ascending order?

(A) $0.7, \frac{3}{4}, 81\%, \frac{17}{20}$ (B) $0.7, \frac{17}{20}, \frac{3}{4}, 81\%$ (C) $\frac{3}{4}, 81\%, 0.7, \frac{17}{20}$ (D) $\frac{17}{20}, 0.7, \frac{3}{4}, 81\%$ — 1

10 The population of a town in 2013 was 4700. In 2014 the population increased by 0.6% and then in 2015 it decreased by 1.3%. Which calculation is used to find the population at the beginning of 2016?

(A) 4700 × 1.006 × 0.87 (B) 4700 × 1.06 × 0.87

(C) 4700 × 1.06 × 0.987 (D) 4700 × 1.006 × 0.987 — 1

Total marks — /10

Fractions, Decimals, Percentages and Ratios

LEVEL 1 TEST — PART A

Time allowed: 40 minutes

Marks

1 What is the value of $75 \div 11.76$, correct to three decimal places?

(A) 6.377 (B) 6.378 (C) 6.387 (D) 6.388 — 1

2 Which of these is 0.8749, correct to two significant figures?

(A) 0.8 (B) 0.9 (C) 0.87 (D) 0.88 — 1

3 Use your calculator to rewrite $\frac{23}{99}$ as a decimal.

(A) 0.2193 (B) 0.2399 (C) 0.23 (D) $0.\dot{2}\dot{3}$ — 1

4 Calculate $\sqrt{8.183}$, leaving your answer correct to three significant figures.

(A) 2.86 (B) 2.87 (C) 2.860 (D) 2.861 — 1

5 In the last census, the population of a city was recorded as 192 638. What is this number written correct to three significant figures?

(A) 192 (B) 193 (C) 192 000 (D) 193 000 — 1

6 Which of these is the same as $0.17\dot{3}$?

(A) 0.173 173 173... (B) 0.333 333 333... (C) 0.173 333 333... (D) 173.173 173 173... — 1

7 Simone used her calculator to evaluate $\frac{3}{5}+\frac{1}{3}$. What was her answer?

(A) $\frac{1}{5}$ (B) $\frac{4}{8}$ (C) $\frac{14}{15}$ (D) $\frac{2}{6}$ — 1

8 Find the value of $(3.4 - 2.8)^2$.

(A) –4.44 (B) 0.36 (C) 0.12 (D) 0.6 — 1

9 Which of these is the largest?

(A) 2 (B) 3% (C) 0.04 (D) 24% — 1

10 Which of these is equal to 8% of $2000?

(A) $160 (B) $1600 (C) $2160 (D) $3600 — 1

11 Remi scored 6 out of 10 in a Science quiz. Which of these is his result as a percentage?

(A) 3% (B) 6% (C) 30% (D) 60% — 1

12 Simplify 40 cents : $2.

(A) 1 : 5 (B) 1 : 4 (C) 1 : 2 (D) 20 : 1 — 1

13 On 1 January 2016 it was estimated that 671 980 000 people did not have access to a safe source of drinking water. What is this number written in scientific notation?

(A) 6.7198×10^6 (B) 6.7198×10^7 (C) 6.7198×10^8 (D) 6.7198×10^9 — 1

Fractions, Decimals, Percentages and Ratios

LEVEL 1 TEST PART B

Write answers and working in the space provided.

Marks

14 Calculate, leaving your answer correct to two decimal places:

a $\sqrt{447.29}$ ________ **b** $53 \div 9$ ________ **c** $\frac{14.2}{8.3}$ ________ 3

15 Simplify the following ratios:

a 10 : 100 ________ **b** 25 : 15 ________ **c** 4 : 9 ________ 3

16 Convert to decimals:

a $\frac{3}{4}$ ________ **b** 18% ________ **c** 4% ________ 3

17 Convert to percentages:

a $\frac{2}{25}$ ________ **b** 0.8 ________ **c** 0.05 ________ 3

18 Convert to fractions:

a 0.6 ________ **b** 0.02 ________ **c** 24% ________ 3

19 Rewrite these numbers in scientific notation:

a 860 ________ **b** 4590 ________

c 3 700 000 ________ 3

20 These numbers are written in scientific notation. Rewrite each number as a basic numeral:

a 2.675×10^2 ________ **b** 5.79329×10^3 ________

c 7.2982×10^4 ________ 3

21 Evaluate, writing your answer to the nearest whole number:

a 12.97×4.81 ________ **b** $19.539 \div 2.9$ ________ 2

22 Find:

a $\frac{3}{4}$ of 732 ________ **b** $\frac{31}{100}$ of 400 ________ 2

23 Bryce scored the following results in tests. Change each of them to percentages:

a Science: $\frac{32}{40}$ ________ **b** History: $\frac{51}{60}$ ________

________ ________ 2

Total marks /40

Fractions, Decimals, Percentages and Ratios

LEVEL 2 TEST — PART A

Time allowed: 40 minutes

Marks

1 Which of these is 49 800 expressed in scientific notation?

(A) 49.8×10^3 (B) 4.98×10^3 (C) 4.98×10^4 (D) 4.98×10^5 — 1

2 Round off 37 297 to two significant figures.

(A) 37 000 (B) 38 000 (C) 37 200 (D) 37 300 — 1

3 A number has been rewritten as 860, correct to two significant figures. Which of these could have been the original number?

(A) 86 (B) 850 (C) 854 (D) 864 — 1

4 Calculate $\frac{19.601}{3.855}$, correct to three significant figures.

(A) 5.08 (B) 5.09 (C) 5.84 (D) 5.85 — 1

5 When a group of hikers arrived at Hunter Springs the temperature was –2.4 °C. After 4 hours the temperature was –6.1 °C. Which of these describes the change in temperature?

(A) fell by 3.7 °C (B) fell by 4.3 °C (C) rose by 3.7 °C (D) rose by 4.3 °C — 1

6 Megan increased 1800 by 12%. Which of these is her result?

(A) 1812 (B) 2016 (C) 2024 (D) 2116 — 1

7 What is the difference between 3×10^4 and 5×10^3?

(A) 10 (B) 20 (C) 2.5×10^3 (D) 2.5×10^4 — 1

8 Which of these is 740 372, written in scientific notation, correct to three significant figures?

(A) 7.40×10^5 (B) 7.4037×10^5 (C) 7.41×10^4 (D) 7.40×10^6 — 1

9 Which of these is closest to 0.007 55?

(A) 0.007 (B) 0.0074 (C) 0.0076 (D) 0.007 35 — 1

10 What is the average of –12, 6 and 9?

(A) 1 (B) 6 (C) 3 (D) –3 — 1

11 Which of these is closest to 1.63?

(A) $\sqrt{2.64}$ (B) $\frac{3.96}{2.4}$ (C) $\frac{1}{0.63}$ (D) $\sqrt{\frac{1}{0.376}}$ — 1

12 One morning Maya leaves Canberra and by lunchtime has driven 40% of the total distance to her grandmother's home in Adelaide. If the distance from Canberra to Adelaide is 1160 km, how far has she yet to travel?

(A) 464 km (B) 484 km (C) 676 km (D) 696 km — 1

13 Simplify $\frac{1}{3} : 1\frac{1}{2}$.

(A) 2 : 9 (B) 2 : 5 (C) 1 : 2 (D) 2 : 3 — 1

Fractions, Decimals, Percentages and Ratios

LEVEL 2 TEST — PART B

Write answers and working in the space provided.

Marks

14 Calculate, leaving your answer correct to two significant figures:

a $\dfrac{19.43 - 12.09}{3.981}$ ________ **b** $\sqrt{\dfrac{14.2}{9}}$ ________

c $\dfrac{2}{\sqrt{19.673}}$ ________ 3

15 Express the following in scientific notation:

a 1270 ________ **b** 23 908 ________ **c** 2 506 000 ________ 3

16 Simplify the following ratios:

a \$6 : 30 cents ________ **b** 20 seconds : 3 minutes ________

c 5 cm : 10 m ________ **d** 0.001 : 0.000 001 ________ 4

17 Divide \$720 in the following ratios:

a 2 : 3 ________ **b** 4 : 3 : 2 ________ 4

18 Evaluate, leaving your answer in index form:

a $3^2 \times 3^4$ ________ **b** $5^6 \div 5$ ________ **c** $(2^3)^2$ ________ 3

19 a Potatoes cost \$2.49/kg. What mass of potatoes can be purchased for \$10? ________ 1

b Simon cycles 21 km in half an hour. What is his average speed in km/h? ________ 1

20 Express:

a 10 m/s in km/h ________ **b** 80 km/h in m/s ________ 4

21 a The angles of a triangle are in the ratio 2 : 3 : 4. What is the size of the smallest angle? ________ 2

b Water is dripping at the rate of 50 mL/min. How much water, in kilolitres, is wasted in one week? ________ 2

Total marks /40

Fractions, Decimals, Percentages and Ratios

LEVEL 3 TEST — PART A

Time allowed: 40 minutes

Marks

1 Evaluate $\frac{2.6+1.901}{\sqrt{3.81}}$, correct to two decimal places. [1]

(A) 1.76 (B) 2.31 (C) 2.74 (D) 3.57

2 Which of the following equals $0.\dot{5}$? [1]

(A) $\frac{1}{5}$ (B) $\frac{1}{2}$ (C) $\frac{5}{9}$ (D) $\frac{5}{99}$

3 What is the value of $(3.081 \times 10^3)^4$, correct to two significant figures? [1]

(A) 1.3×10^7 (B) 3.9×10^{12} (C) 9.5×10^{12} (D) 9.0×10^{13}

4 Last week Don read half the pages of his novel. Last night he read one-quarter of the remaining pages. If there are still 60 pages yet to read, what is the total number of pages in the novel? [1]

(A) 160 (B) 180 (C) 200 (D) 240

5 Evaluate $\sqrt{\frac{9.28}{\pi}}$, correct to four significant figures. [1]

(A) 0.9697 (B) 0.970 (C) 1.719 (D) 1.718

6 Which of these is closest to $\frac{1.8}{0.1+0.6^2}$? [1]

(A) 3.7 (B) 3.9 (C) 5.1 (D) 6.6

7 Increase \$520 by 18% and then decrease this amount by 13%. What is the result? [1]

(A) \$494 (B) \$525 (C) \$533.83 (D) \$546

8 Which of these is the same as $100\ \text{m}^2 : 1$ ha? [1]

(A) 1 : 100 (B) 1 : 10 (C) 10 : 1 (D) 100 : 1

9 A car uses fuel at the rate of 6.9 L per 100 km. How far will the car travel using 50 L? [1]

(A) 345 km (B) 690 km (C) 710 km (D) 725 km

10 It takes Laura 40 minutes to clean her room. When her mother cleans Laura's room it only takes 15 minutes. Which of these is closest to the time it would take to clean the room if they both worked together? [1]

(A) 11 minutes (B) 12 minutes (C) 13 minutes (D) 14 minutes

A bag contains green, red and blue balls. There are 36 balls in the bag, the ratio of green balls to red balls is 3 : 2 and the ratio of red balls to blue balls is 1 : 2. This information is used to answer questions 11, 12 and 13.

11 What is the ratio of green balls to blue balls? [1]

(A) 3 : 4 (B) 3 : 2 (C) 3 : 1 (D) 1 : 3

12 How many blue balls are in the bag? [1]

(A) 6 (B) 8 (C) 12 (D) 16

13 If another four red balls are added to the bag, what is the ratio of red balls to green balls? [1]

(A) 1 : 1 (B) 1 : 2 (C) 2 : 5 (D) 3 : 1

Fractions, Decimals, Percentages and Ratios

LEVEL 3 TEST — PART B

Write answers and working in the space provided.

Marks

14 Evaluate, leaving your answer in scientific notation, correct to three significant figures:

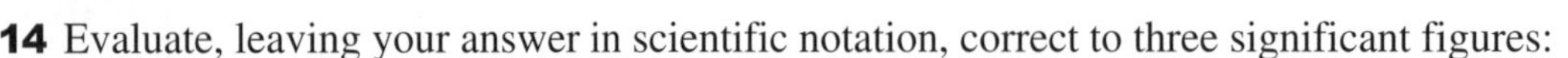

a $\dfrac{2.06 \times 10^8}{1.9 \times 10^6}$ ______ **b** $\sqrt{3.96 \times 10^7}$ ______

c $(5.2 \times 10^5) \times (2.08 \times 10^3)$ ______ **d** $(9.12 \times 10^4)^7$ ______ [4]

15 Express the following in the form of $\frac{p}{q}$, where p and q are integers:

a $0.\dot{6}\dot{9}$ ______ **b** $0.2\dot{6}$ ______ [4]

16 Which single percentage increase is the same as:

a two consecutive increases of 20%? **b** an increase of 20% and then a decrease of 10%? [4]

17 What is the middle of:

a $\frac{1}{3}$ and $\frac{1}{5}$? **b** $\frac{2}{5}$ and $1\frac{1}{2}$? **c** $-\frac{1}{2}$ and 0.7? [6]

18 Evaluate, leaving your answer correct to three significant figures:

a $\sqrt[3]{\dfrac{18.6 - 12.8}{2.1}}$ ______ **b** $\dfrac{1}{4.7 - \sqrt{18.71}}$ ______

c $\sqrt{\dfrac{1.89}{0.38 + 11.6}}$ ______ [3]

19 a The landmass of Australia is 8.468 million km^2. If the population of the earth is 7.3 billion, how much space would each person have if they all lived in Australia? Give your answer in square metres. [2]

b Quentin won a basket of Easter eggs in a raffle. He gave one-third to his family, half of the remainder to his friends, kept six eggs for himself and gave the other two eggs to his neighbour. How many eggs were originally in the basket? [2]

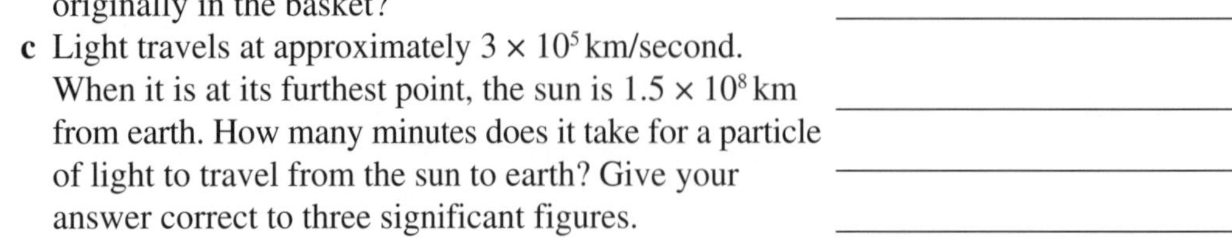

c Light travels at approximately 3×10^5 km/second. When it is at its furthest point, the sun is 1.5×10^8 km from earth. How many minutes does it take for a particle of light to travel from the sun to earth? Give your answer correct to three significant figures. [2]

Total marks

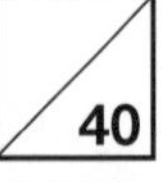

/40

Financial Mathematics

PRETEST — LEVEL 1

Time allowed: 10 minutes — **Total marks: 10**

	Marks
1 Jack bought a quad bike for $920 and later sold it for a profit of $175. What was his selling price? (A) $745 (B) $995 (C) $1075 (D) $1095	1
2 What is 20% of $720? (A) $14.40 (B) $144 (C) $734.40 (D) $864	1
3 A car is discounted by $2500 from an original price of $30 000. What is the car's new price? (A) $500 (B) $7500 (C) $27 500 (D) $32 500	1
4 What is 8% written as a decimal? (A) 0.008 (B) 0.08 (C) 0.8 (D) 0.108	1
5 What is the value of $0.175 \times 1600 \times 4$? (A) 920 (B) 1120 (C) 1420 (D) 1480	1
6 How many per cent is in 'one whole'? (A) 1 (B) 10 (C) 100 (D) 1000	1
7 Todd bought a wardrobe for $360 and later sold it to make a loss of $40. At what price did Todd sell the wardrobe? (A) $320 (B) $364 (C) $400 (D) $480	1
8 A furniture shop held a storewide sale of 10% discount on all stock. Dayne purchased a lounge suite originally priced at $1200. Which of these amounts represents the amount of discount received by Dayne? (A) $12 (B) $120 (C) $1080 (D) $1290	1
9 Each week Katie earns $980 working as a shop assistant. How much will she be paid if she works a total of 42 weeks? (A) $41 160 (B) $41 260 (C) $41 360 (D) $41 460	1
10 Sandra places pamphlets into letterboxes. She is paid 6 cents for each bundle of pamphlets she delivers. How much will Sandra receive if she delivers 860 bundles of pamphlets? (A) $5.16 (B) $8.66 (C) $51.60 (D) $86.60	1

Total marks /10

Financial Mathematics

PRETEST

LEVELS 2 AND 3

Time allowed: 10 minutes **Total marks: 10**

Marks

1 What is $8\frac{1}{2}\%$ of \$400?

(A) \$32.48 (B) \$32.80 (C) \$34 (D) \$340 1

2 The price of a new tablet is to increase by 6%. If the tablet originally costs \$550, what is the price of the new product?

(A) \$517 (B) \$544 (C) \$556 (D) \$583 1

3 A certain country presently sets the Goods and Services Tax (GST) at 10%. If a phone costs \$630 without GST, what is its price when GST is included?

(A) \$640 (B) \$665 (C) \$689 (D) \$693 1

4 Which of these is the best buy for packets of toilet paper?

(A) 6 rolls for \$3.20 (B) 8 rolls for \$4.90 (C) 12 rolls for \$6.60 (D) 18 rolls for \$9.45 1

5 The price of a leaf blower is discounted by 15%. If it was originally priced at \$220, what is the new price?

(A) \$187 (B) \$195 (C) \$205 (D) \$253 1

6 Increase \$380 by 35%.

(A) \$133 (B) \$415 (C) \$513 (D) \$525 1

7 A coffee machine priced at \$680 is discounted by \$136. What is the discount expressed as a percentage?

(A) 12% (B) 20% (C) 24% (D) 36% 1

8 Grace bought some jewellery online for \$260 and later sold it for \$221. What was the loss as a percentage of the cost price?

(A) 15% (B) 24% (C) 76% (D) 85% 1

9 Decrease \$100 by 20% and then increase this amount by 20%. What is the final amount?

(A) \$80 (B) \$96 (C) \$100 (D) \$104 1

10 Year 9 raised \$2400 as part of the whole school's fundraising week. If this amount represented 16% of the money raised by the school, what was the total fundraising amount?

(A) \$13 800 (B) \$15 000 (C) \$18 600 (D) \$20 000 1

Total marks

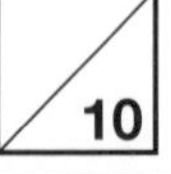

Financial Mathematics

LEVEL 1 TEST — PART A

Time allowed: 40 minutes

Marks

1 William is paid \$1288 each fortnight. Which of these is his weekly wage?
(A) \$92 (B) \$128.80 (C) \$644 (D) \$2576 — 1

2 Simone has a casual job and is paid \$18.65 per hour. What will she be paid if she works 16 hours?
(A) \$34.65 (B) \$298 (C) \$298.40 (D) \$312.35 — 1

3 Darcie receives \$667 for working 23 hours. What is her hourly rate?
(A) \$21 (B) \$29 (C) \$31 (D) \$39 — 1

4 Hayley is paid an hourly rate of \$18.40. How much will she be paid if she works for 8 hours at time-and-a-half?
(A) \$174.80 (B) \$220.80 (C) \$264.40 (D) \$294.40 — 1

5 What is the amount of simple interest earned on a deposit of \$500 at 4% p.a. over 7 years?
(A) \$127 (B) \$140 (C) \$1400 (D) \$3640 — 1

6 To work out his net pay (take home pay), Mick used this rule: *Net pay = Gross pay – Deductions*
What is his net pay if his gross pay is \$2400 and he had a total of \$310 worth of deductions?
(A) \$1990 (B) \$2090 (C) \$2190 (D) \$2710 — 1

7 Alana is paid \$45 for every 1000 brochures she delivers. What amount will she be paid if she delivers 3000 brochures?
(A) \$115 (B) \$125 (C) \$135 (D) \$145 — 1

8 Ian is paid double-time when he works on Sundays. If his normal pay is \$21.30 per hour, how much is he paid if he works 6 hours on a Sunday?
(A) \$174.04 (B) \$174.40 (C) \$255.06 (D) \$255.60 — 1

9 Phoebe is presently paid \$76 890 p.a. Next year she is to receive a pay rise of 2.3%. What will be the increase in her salary?
(A) \$1560.87 (B) \$1768.47 (C) \$17 684.70 (D) \$78 658.47 — 1

10 Each week Dante contributes \$85 into his superannuation fund. What is the total of the contribution in a year?
(A) \$4250 (B) \$4400 (C) \$4410 (D) \$4420 — 1

11 Hope pays her car loan by monthly instalments of \$340. How much will she pay in 3 years?
(A) \$1020 (B) \$12 240 (C) \$12 580 (D) \$16 840 — 1

12 Which of these is \$7800 increased by $2\frac{1}{2}\%$?
(A) \$7995 (B) \$8050 (C) \$9750 (D) \$11 300 — 1

13 Which of these is the largest amount?
(A) \$1280/week (B) \$2310/fortnight (C) \$5100/month (D) \$63 480/year — 1

Financial Mathematics

LEVEL 1 TEST

PART B

Write answers and working in the space provided.

Marks

14 Owen is paid an hourly rate of \$19.30. What is he paid if he works:

a 35 hours at normal pay? **b** 12 hours at time-and-a-half? **c** 5 hours at double-time?

6

15 Rachel is paid a weekly wage of \$675. What will she be paid if she works for:

a 18 weeks? **b** 5 fortnights

4

16 Complete the tables:

a

Fortnightly pay	Annual pay
\$2453	
	\$103 610

2

b

Annual pay	Monthly pay
\$96 852	
	\$7530

2

17 Use the simple interest formula, $I = Prn$, to find the amount of simple interest if:

a \$600 is invested at 4% p.a. for 3 years **b** \$25 000 is invested at 5% p.a. for 7 years

4

18 Edward earns \$1260 per week and is drawing up a budget. He plans to save one-quarter of his pay.

a How much will he save each week? **b** How much will he save in 12 weeks?

4

19 Charlotte receives 4 weeks holiday loading at 17.5% of her normal pay each year. If her normal pay is \$1040 per week, find the:

a amount of holiday loading she will receive **b** total amount she will receive for the 4 weeks

4

20 Complete the table:

Gross pay	Deductions	Net pay
\$67 489		\$61 896

1

Total marks 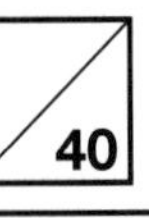 40

Financial Mathematics

LEVEL 2 TEST PART A

Time allowed: 40 minutes

Marks

1 What is the amount of simple interest earned if \$2400 is invested for 8 years at 4.5% p.a.?

(A) \$791.04 (B) \$864 (C) \$3264 (D) \$8640 — 1

2 Nikki's usual pay rate is \$22.35 per hour. What will she be paid for working 28 hours normal time and 8 hours at time-and-a-half?

(A) \$637.80 (B) \$804.60 (C) \$838.13 (D) \$894.00 — 1

3 Corinne receives 4 weeks holiday loading at the rate of 17.5%. What will be the amount of her holiday loading if she is paid \$1280 per week?

(A) \$448 (B) \$896 (C) \$1504 (D) \$6016 — 1

4 Liam's employer contributes into a superannuation fund an amount that equals 9.5% of Liam's fortnightly pay. What will be the contribution each fortnight, to the nearest dollar, if Liam is paid \$76 400 p.a.?

(A) \$279 (B) \$363 (C) \$726 (D) \$7258 — 1

5 On retirement, Michael receives 58.2% of his finishing salary of \$96 890 each year. Which of these is closest to the amount he will receive each fortnight?

(A) \$2169 (B) \$2316 (C) \$2616 (D) \$56 390 — 1

6 Which of these is the best buy?

(A) \$3.60 for 250 mL (B) \$4.50 for 400 mL (C) \$6.20 for 600 mL (D) \$7.40 for 680 mL — 1

7 At a 15% discount sale, the price of an outdoor setting drops by \$180. What was the original price of the furniture?

(A) \$27 (B) \$207 (C) \$1020 (D) \$1200 — 1

8 A bank charges an interest rate of 18.75% p.a. for a credit card. Which of these is closest to the daily interest rate?

(A) 0.005 137% (B) 0.051 37% (C) 0.5137% (D) 5.137% — 1

9 Amelia borrows \$5000 for a holiday. She repays the amount in monthly repayments of \$176 over 3 years. What amount of interest will be paid?

(A) \$1284 (B) \$1336 (C) \$2640 (D) \$6336 — 1

10 Sam is to receive a 1.8% pay rise which means she will receive an extra \$17.37 per week. What will be Sam's new annual salary?

(A) \$903.24 (B) \$50 180 (C) \$51 047.24 (D) \$51 083.24 — 1

11 What amount of GST (at 10%) is included in the price of a car repair costing \$869?

(A) \$79 (B) \$86.09 (C) \$86.90 (D) \$782.10 — 1

12 Audrey works from 8:30 am to 12:15 pm and is paid \$21.20 per hour. Find her pay.

(A) \$68.90 (B) \$73.14 (C) \$79.50 (D) \$90.10 — 1

13 Which of these is the best pay?

(A) \$26/hour normal (B) \$17/hour time-and-a-half

(C) \$13.20/hour double-time (D) \$10.17/hour double-time-and-a-half — 1

Financial Mathematics

LEVEL 2 TEST — PART B

Write answers and working in the space provided.

Marks

The tax table is used to answer questions 14 and 15.

Taxable income	Tax on this income
\$0–\$18 200	Nil
\$18 201–\$37 000	19c for each \$1 over \$18 200
\$37 001–\$87 000	\$3572 plus 32.5c for each \$1 over \$37 000
\$87 001–\$180 000	\$19 822 plus 37c for each \$1 over \$87 000
\$180 001 and above	\$54 232 plus 47c for each \$1 over \$180 000

14 Find the tax payable on a taxable income of:

a \$21 650 ______________ **b** \$37 902 ______________ **c** \$102 652 ______________

6

15 Ava has a gross income of \$87 420 and tax deductions of \$11 438. Find:

a her taxable income

b the amount of tax to be paid

3

16 In a half-yearly clearance sale a department store discounted a series of products. The table below shows five items. Complete the table:

Product	Old price	Discount %	Discount	New price
a	\$640	15%		
b	\$500		\$100	
c	\$720			\$612
d		25%	\$60	
e			\$40	\$360

10

17 Find the simple interest on:

a \$9800 at 5.25% p.a. for 6 years

b \$6860 at 0.4% per month for 2 years

4

18 Sam wants to buy a jet-ski priced at \$12 800. He pays a deposit of 20% and the balance is paid by 24 monthly repayments of \$480 each.

a Find the total cost of the jet-ski.

b Find the annual interest rate charged.

4

Total marks

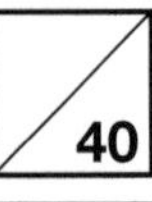

Financial Mathematics

LEVEL 3 TEST PART A

Time allowed: 40 minutes

Marks

1 What is the amount of simple interest earned if $\$p$ is invested for k months at $r\%$ p.a.?

(A) $\frac{prk}{1200}$ (B) $\frac{prk}{100}$ (C) $\frac{12prk}{100}$ (D) $\frac{prk}{12}$ 1

2 A shop had a '$x\%$ off everything' discount sale. If the price of a coffee table dropped from $360 to $288, what is the value of x?

(A) 16 (B) 18 (C) 20 (D) 25 1

3 Amelia borrows $7500 and repays the amount plus interest. If she makes 24 monthly repayments of $350, what is the annual rate of simple interest?

(A) 3% (B) 4% (C) 5% (D) 6% 1

4 The price of a mattress, including GST of 10%, is $957. If the GST is to rise from 10% to 15%, what will be the new price of the mattress?

(A) $972 (B) $982 (C) $1000.50 (D) $1100.55 1

5 What amount is invested to earn simple interest of $336 over 3 years at 4% p.a.?

(A) $376.32 (B) $2800 (C) $2920 (D) $3040 1

6 Using the tax table on page 14, what amount of tax is payable on an income of $112 671?

(A) $9498.27 (B) $25 935.27 (C) $29 320.27 (D) $41 688.27 1

7 What is the compound interest on $12 000 invested for 5 years at 6.3% p.a.?

(A) $3780 (B) $4287.24 (C) $15 780 (D) $16 287.24 1

8 A car, purchased for $43 500, loses value at an annual rate of 18%. Which of these is the depreciated value of the car after 8 years, to the nearest dollar?

(A) $8892 (B) $19 140 (C) $19 634 (D) $29 100 1

9 Jordie invests $10 000 in an account earning 5% p.a. simple interest. In how many years will he double his money?

(A) 2 (B) 20 (C) 40 (D) 100 1

10 Using the tax table on page 14, what is Emily's taxable income if she paid $19 822.74 tax?

(A) $87 001 (B) $87 002 (C) $87 003 (D) $87 004 1

11 Ryan is paid a base amount of $720 per week plus 4% on any sales above $12 000. If last week he was paid a total of $1350, what was the total value of sales?

(A) $2550 (B) $24 960 (C) $26 280 (D) $27 750 1

12 In one week James worked 28 hours normal time and 8 hours at time-and-a-half. If James was paid $850, how much will he be paid to work 5 hours at double-time?

(A) $193.18 (B) $212.50 (C) $237.30 (D) $241.20 1

13 Noah invests $5000 at 8% p.a. compound interest for 12 years. If he had invested the same amount in an account earning 8% p.a. simple interest, how long, to the nearest year, would it take to earn the same amount?

(A) 12 years (B) 15 years (C) 17 years (D) 19 years 1

Financial Mathematics

LEVEL 3 TEST — PART B

Write answers and working in the space provided.

Marks

14 Use the tax table on page 14 to find Mason's taxable income if he paid tax of:

a \$19 367 ______ **b** \$40 172 ______ [4]

15 What amount of compound interest is earned after investing:

a \$18 000 for 5 years at 7.3% p.a.? **b** \$12 500 for 2 years at 6% p.a.? [4]

16 Find the size of each monthly instalment.

a A boat is priced at \$68 900. A trade-in boat valued at \$12 000 is accepted as a deposit, and the remainder is paid in monthly repayments at 5.6% p.a. simple interest for 4 years. [3]

b A car is purchased for \$56 300. A deposit of 20% is paid and the balance paid in 36 monthly instalments. The loan attracts a simple interest rate of 8% p.a. [3]

17 A retailer sold a bedroom suite for \$2799.50. The sale price included 10% GST and the retailer made a 35% profit on his cost price. Find:

a the pre-GST selling price **b** the amount of profit, to the nearest dollar [4]

18 In May 2010 a rare coin was valued at \$22 000.
For the following 2 years it appreciated in value by 8% p.a. and then for 3 years it depreciated in value by 5% p.a.

a What was its value in May 2013? ______ [3]

b How much did it drop in value in the fifth year? ______ [3]

19 Mark purchases an \$80 000 caravan under the following terms:

- 10% deposit
- balance repaid in monthly instalments of \$2360
- 3-year loan.

What annual simple interest rate will be charged? [3]

Total marks

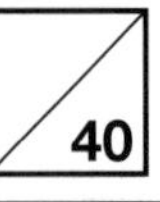

Algebraic Techniques

PRETEST — LEVEL 1

Time allowed: 10 minutes — **Total marks: 10**

Marks

1 Simplify $3x + 5x$.

(A) $8x$ (B) $8x^2$ (C) x^8 (D) $15x^2$ — 1

2 What is $5y - y$?

(A) 0 (B) 4 (C) $4y$ (D) 5 — 1

3 Simplify $6a \times 3$.

(A) 18 (B) $9a$ (C) $18a$ (D) $18a^2$ — 1

4 Which of these is the same as $p + p + p + p$?

(A) p^4 (B) $4p$ (C) $2p^2$ (D) $4p^4$ — 1

5 Simplify $10x \div 5$.

(A) 2 (B) $2x$ (C) 5 (D) $5x$ — 1

6 Simplify $4y - 7y$.

(A) -11 (B) $-11y$ (C) -3 (D) $-3y$ — 1

7 A box contains three black balls and two white balls.
A second box contains six black balls and four white balls.

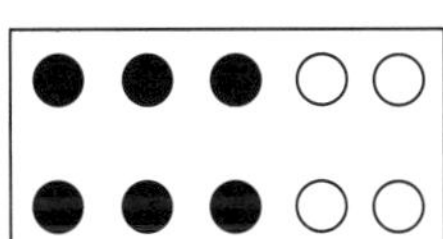

How many balls are there altogether?

(A) six black balls and nine white balls (B) nine black balls and six white balls

(C) nine black balls and nine white balls (D) 15 black balls and 15 white balls — 1

8 Which of these is equal to $5a + 3b - 4a + b$?

(A) $a + 4b$ (B) $a + 2b$ (C) $5ab$ (D) $a^2 + 4b^2$ — 1

9 Which of these is the same as $\frac{2}{3}$?

(A) $\frac{1}{3}$ (B) $\frac{4}{9}$ (C) $\frac{4}{6}$ (D) $\frac{3}{2}$ — 1

10 Evaluate $\frac{1}{2} \times \frac{2}{3}$.

(A) $\frac{1}{3}$ (B) $\frac{4}{9}$ (C) $\frac{3}{5}$ (D) 6 — 1

Total marks /10

Algebraic Techniques

PRETEST — LEVELS 2 AND 3

Time allowed: 10 minutes — **Total marks: 10**

Marks

1 What is the sum of t, $2t$ and $3t$?

(A) $5t$ (B) $6t$ (C) $5t^3$ (D) $6t^3$ — 1

2 What is the missing expression?

$3a \times$ ▭ $= 12a$

(A) 4 (B) $4a$ (C) 9 (D) $9a$ — 1

3 Simplify $5x \times (-4y)$.

(A) $-xy$ (B) xy (C) $20xy$ (D) $-20xy$ — 1

4 Simplify $(-3a)^2$.

(A) $-9a^2$ (B) $-6a^2$ (C) $6a^2$ (D) $9a^2$ — 1

5 What is the perimeter of the triangle?

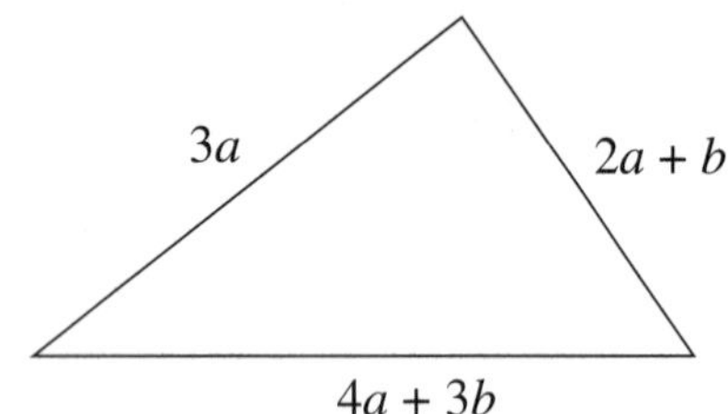

(A) $a + b$ (B) $3a + 3b$ (C) $9a + 4b$ (D) $9a + 5b$ — 1

6 Evaluate $\frac{1}{2} + \frac{1}{3}$.

(A) $\frac{1}{6}$ (B) $\frac{1}{5}$ (C) $\frac{2}{5}$ (D) $\frac{5}{6}$ — 1

7 What is the value of $\frac{7}{10} - \frac{2}{10}$?

(A) 0 (B) $\frac{1}{5}$ (C) $\frac{1}{2}$ (D) 5 — 1

8 What is the value of $2(3 + 1)$?

(A) 6 (B) 8 (C) 10 (D) 12 — 1

9 Which of these is a factor of both 4 and 6?

(A) 2 (B) 4 (C) 6 (D) 24 — 1

10 Which of the following is **not** a factor of $8a$?

(A) $4a$ (B) $8a$ (C) 1 (D) $a + 8$ — 1

Total marks 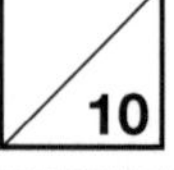 /10

Algebraic Techniques

LEVEL 1 TEST — PART A

Time allowed: 40 minutes

Marks

1 Simplify $\frac{5a}{5}$.
Ⓐ a Ⓑ $5a^2$ Ⓒ $25a$ Ⓓ $25a^2$ [1]

2 Which of these is equal to $3p + 2q + p + 4q$?
Ⓐ $9pq$ Ⓑ $10pq$ Ⓒ $4p + 6q$ Ⓓ $4p^2 + 6q^2$ [1]

3 Expand $2(m + 4)$.
Ⓐ $2m + 4$ Ⓑ $2m + 6$ Ⓒ $8m$ Ⓓ $2m + 8$ [1]

4 Simplify $\frac{a}{5} + \frac{2a}{5}$.
Ⓐ $\frac{3a}{5}$ Ⓑ $\frac{3a}{10}$ Ⓒ $\frac{2a}{25}$ Ⓓ $\frac{2a^2}{25}$ [1]

5 Simplify $\frac{p}{3} \times \frac{p}{2}$.
Ⓐ $\frac{p^2}{5}$ Ⓑ $\frac{p^2}{6}$ Ⓒ $\frac{1}{5}$ Ⓓ $\frac{1}{6}$ [1]

6 Factorise $5a - 15$.
Ⓐ $5(a - 3)$ Ⓑ $3(a - 5)$ Ⓒ $15(a - 1)$ Ⓓ $15(a - 3)$ [1]

7 Which expression is $p \times p \times q \times q \times q$?
Ⓐ $2p + 3q$ Ⓑ p^2q^3 Ⓒ $5pq$ Ⓓ $p^2 + q^3$ [1]

8 Simplify $4m - 2m - 2m$?
Ⓐ m Ⓑ 0 Ⓒ $8m$ Ⓓ $-2m$ [1]

9 Expand $3(2y - 5)$.
Ⓐ $5y - 8$ Ⓑ $5y - 2$ Ⓒ $6y - 2$ Ⓓ $6y - 15$ [1]

10 Simplify $\frac{6x}{x}$.
Ⓐ $5x$ Ⓑ 6 Ⓒ 5 Ⓓ x^5 [1]

11 Simplify $3a - 2a^2 + a$.
Ⓐ $4a - 2a^2$ Ⓑ $2a - 2a^2$ Ⓒ 0 Ⓓ $2a^2$ [1]

The rectangle below is used to answer questions 12 and 13.

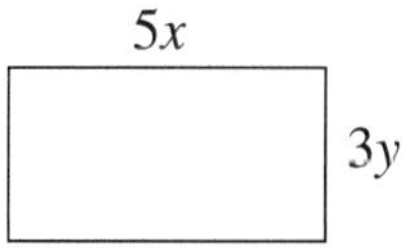

12 What is the area of the rectangle?
Ⓐ $8xy$ Ⓑ $15xy^2$ Ⓒ $5x + 3y$ Ⓓ $15xy$ [1]

13 What is the perimeter of the rectangle?
Ⓐ $5x + 3y$ Ⓑ $25x + 9y$ Ⓒ $10x + 6y$ Ⓓ $10x^2 + 6y^2$ [1]

Algebraic Techniques

LEVEL 1 TEST — PART B

Write answers and working in the space provided.

Marks

14 Simplify:

a $5p + 4 - p - 3$ b $11 + 6d + 4 - 2d$ c $5x - 8x$ 3

15 Simplify:

a $4y \times 2$ b $10ab \times 3$ c $(-4a) \times (-3a)$ 3

16 Simplify:

a $6a \div 3$ b $4ab \div 4a$ c $(-6p) \div 6$ 3

17 Simplify:

a $\frac{4y}{2}$ b $\frac{10x}{x}$ c $\frac{6p}{8}$ 3

18 Expand:

a $5(y + 1)$ b $3(m + n)$ c $2(3a - 5)$ 3

19 Simplify:

a $3p + 4 \times p$ b $10ab - 3a \times 4b$ 4

20 Simplify:

a $\frac{4t}{3} + \frac{t}{3}$ b $\frac{7a}{5} - \frac{3a}{5}$ c $\frac{4}{x} + \frac{3}{x}$ 3

21 Factorise:

a $3y - 12$ b $4a + 20$ c $6x - 8$ 3

22 Expand and simplify $3(p + 4) + 5$. 2

Total marks

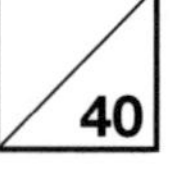

Algebraic Techniques

LEVEL 2 TEST — PART A

Time allowed: 40 minutes

Marks

1 Simplify $5x - x^2 + 2x^2 - 6x$.

(A) $-x$ (B) $-x - 3x^2$ (C) $-x - x^2$ (D) $-x + x^2$ — 1

2 Simplify $\frac{8}{4x}$.

(A) $2x$ (B) $\frac{4}{x}$ (C) $\frac{2}{x}$ (D) $4x$ — 1

3 Which of the following expressions equals $15pq - 5p \times 4q$?

(A) $-40pq$ (B) $-5pq$ (C) $5pq$ (D) $40pq$ — 1

4 Find the perimeter of the triangle:

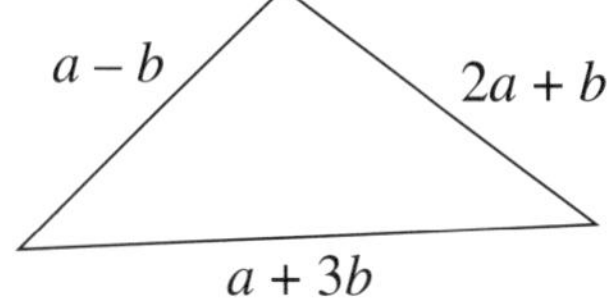

(A) $2a + 3b$ (B) $4a + 3b$ (C) $4a - 3b$ (D) $4a + 4b$ — 1

5 Expand and simplify $4(x - 3) + 2(x - 1)$.

(A) $6x - 4$ (B) $6x - 10$ (C) $6x - 14$ (D) $2x - 14$ — 1

6 Simplify $\frac{5}{2m} + \frac{7}{2m}$.

(A) $6m$ (B) $12m$ (C) $\frac{6}{m}$ (D) $\frac{12}{m}$ — 1

7 Which of these is the same as $\frac{x}{5} \times \frac{10}{3x}$?

(A) $\frac{2}{3}$ (B) $\frac{2}{3x}$ (C) $\frac{2x}{3}$ (D) $\frac{10x}{3}$ — 1

8 Expand and simplify $(p + 5)(p - 2)$.

(A) $p^2 + 3p - 10$ (B) $p^2 - 3p - 10$ (C) $p^2 - 3p + 10$ (D) $p^2 + 3p + 10$ — 1

9 Factorise fully $12ab - 6a^2$.

(A) $3(ab - 2a^2)$ (B) $6(ab - a^2)$ (C) $6a^2(2b - a)$ (D) $6a(2b - a)$ — 1

10 Simplify $\frac{3m + 6n}{3}$.

(A) $\frac{m + 2n}{3}$ (B) $m + 6n$ (C) $m + 3n$ (D) $m + 2n$ — 1

11 What is the sum of $(4x - 3y)$ and $(y - x)$?

(A) $3x - 2y$ (B) $5x - 2y$ (C) $5x - 4y$ (D) $3x - 4y$ — 1

12 Factorise $x^2 - x$.

(A) $x(x - 1)$ (B) $x^2(x - 1)$ (C) $x(2x - 1)$ (D) $2(x - 1)$ — 1

13 Expand and simplify $2a - a(a - 1)$.

(A) $-a^2 - a$ (B) $-a^2 - 3a$ (C) $-a^2 + 3a$ (D) $a^2 + a$ — 1

Algebraic Techniques

LEVEL 2 TEST PART B

Write answers and working in the space provided.

Marks

14 Simplify:

a $5ab - 7a - 3ba + 2a$ b $-4m - 2mn + nm - 3m$

2

15 Simplify:

a $\frac{4c}{8cd}$ b $\frac{3m}{4n} \times \frac{2n}{9m}$ c $\left(\frac{2a}{5b}\right)^2$

3

16 Simplify:

a $\frac{2t}{3} + \frac{5t}{9}$ b $\frac{s}{4} \div \frac{s}{2}$

4

17 Expand:

a $-7(2c + d)$ b $b^2(2b - 1)$ c $-(2a - 3b)$

3

18 Expand and simplify:

a $10(a - 1) + 3(a - 4)$ b $2(5y + 2) - (2y - 5)$

4

19 Factorise:

a $-2a - 2$ b $p(2a + b) + 3(2a + b)$

2

20 Expand and simplify:

a $(x + 2)(x + 3)$ b $(a + 3)(a - 4)$ c $(2b - 1)(3b + 5)$

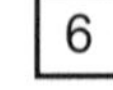

6

21 Simplify $\frac{5p-1}{2} + \frac{4p}{3}$.

3

Total marks

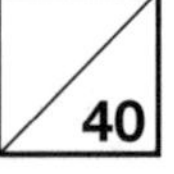

40

Algebraic Techniques

LEVEL 3 TEST — PART A

Time allowed: 40 minutes

Marks

1 Expand $-(3x + 5y - 2)$.

Ⓐ $-3x - 5y + 2$ Ⓑ $3x - 5y + 2$ Ⓒ $-3x - 5y - 2$ Ⓓ $-3x + 5y + 2$ [1]

2 Simplify $\frac{3}{2a} - \frac{8}{3a}$.

Ⓐ 4 Ⓑ $-\frac{7}{6a}$ Ⓒ $-\frac{5}{a}$ Ⓓ $-\frac{1}{6a}$ [1]

3 Which is the difference between $2a - b$ and $b - 2a$?

Ⓐ $4a$ Ⓑ $4a - 2b$ Ⓒ $-2b$ Ⓓ $4a + 2b$ [1]

4 Simplify $\frac{4a}{3b} \times \frac{6}{2a} \div \frac{2}{3b}$.

Ⓐ $\frac{2b}{a}$ Ⓑ $\frac{3b}{2a}$ Ⓒ $\frac{b}{3a}$ Ⓓ 6 [1]

5 Expand and simplify $(4p - 3q)^2$.

Ⓐ $8p^2 - 14pq + 9q^2$ Ⓑ $8p^2 - 24pq + 9q^2$ Ⓒ $16p^2 - 24pq + 9q^2$ Ⓓ $16p^2 - 14pq + 9q^2$ [1]

6 Simplify $2 - \frac{p}{q}$.

Ⓐ $\frac{2q - p}{p}$ Ⓑ $\frac{2p - q}{p}$ Ⓒ $\frac{2q - p}{q}$ Ⓓ $\frac{2p - q}{q}$ [1]

7 Simplify $\left(\frac{1}{a} \times \frac{1}{b}\right) \div \left(\frac{1}{a} \div \frac{1}{b}\right)$.

Ⓐ $\frac{1}{b^2}$ Ⓑ $\frac{a}{b}$ Ⓒ $\frac{b^2}{a}$ Ⓓ 1 [1]

8 Which of these is equal to $\frac{x^2 - 2x}{x^2 - 4}$?

Ⓐ $\frac{x + 2}{x}$ Ⓑ $\frac{x}{2}$ Ⓒ $\frac{x}{x - 2}$ Ⓓ $\frac{x}{x + 2}$ [1]

9 Expand and simplify $(a - b)^2 - (a^2 + b^2)$.

Ⓐ $2a^2 - 2b^2$ Ⓑ $2ab$ Ⓒ $-2ab$ Ⓓ $-2b^2$ [1]

10 What is the area of a square with perimeter $(4n - 4)$ cm?

Ⓐ $(n^2 + 2n + 1)$ cm^2 Ⓑ $(n^2 - 2n + 1)$ cm^2

Ⓒ $(4n^2 - 16n + 16)$ cm^2 Ⓓ $(16n^2 - 64n + 64)$ cm^2 

11 What is the square of $(2 - 3m)$?

Ⓐ $4 - 9m^2$ Ⓑ $4 + 9m^2$ Ⓒ $4 - 6m + 9m^2$ Ⓓ $4 - 12m + 9m^2$ [1]

12 Factorise $ab \quad 4a \quad 3b + 12$.

Ⓐ $(a - 4)(b - 3)$ Ⓑ $(a - 4)(b + 3)$ Ⓒ $(a - 3)(b + 4)$ Ⓓ $(a - 3)(b - 4)$ [1]

13 Simplify $1 - \frac{a}{a - b}$.

Ⓐ $\frac{b}{b - a}$ Ⓑ $\frac{a + b}{a - b}$ Ⓒ $\frac{b - 2a}{a - b}$ Ⓓ $\frac{b}{a - b}$ [1]

LEVEL 3 TEST

PART B

Write answers and working in the space provided.

Marks

14 Simplify:

a $-3a^2 + 2b + 4a^2 - 5b + a^2$

b $xy - 3x^2y + 5xy - 2xy^2$

[2]

15 Expand and simplify:

a $6x + 4(x - 1)$

b $5 - (2 - m) - (3m + 2)$

c $3 - (ab + b) - a(b - 1)$

[6]

16 Simplify:

a $\frac{4x}{3y} \div \left(\frac{2x}{9} \times \frac{3}{y}\right)$

b $\frac{3c+1}{4} - \frac{2c-5}{2}$

c $2 - \frac{4-3p}{5}$

[6]

17 Expand and simplify:

a $(a + b)^2 + (a - b)^2$

b $(x - y)^2 - (x + y)^2$

[4]

18 Expand and simplify:

a $(2p + 3q)^2 - (p - 2q)^2$

b $(2m + n)^2 - 2(m - n)(n - m)$

[6]

19 Simplify $\frac{3x^2 - 3}{1 - x}$.

[3]

Total marks

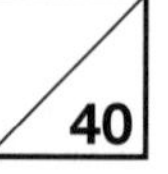

40

Indices

PRETEST — LEVEL 1

Time allowed: 10 minutes — **Total marks: 10**

Marks

1 Which of these is equal to $4a$?

(A) $4 + a$ (B) $a \times a \times a \times a$ (C) $a + a + a + a$ (D) 4^a — 1

2 Which of these is equal to p^5?

(A) $p + p + p + p + p$ (B) $p \times p \times p \times p \times p$ (C) $p + 5$ (D) $p \times 5$ — 1

3 What is the value of -3×-2?

(A) -6 (B) -5 (C) 5 (D) 6 — 1

4 What is the value of $\sqrt{19}$, to the nearest whole?

(A) 364 (B) 4 (C) 5 (D) 10 — 1

5 Which of these is equal to 16?

(A) 8^2 (B) 4^4 (C) 2^4 (D) 1^6 — 1

6 Which of these is **not** a factor of 18?

(A) 1 (B) 3 (C) 9 (D) 12 — 1

7 Simplify $aa + a^2$.

(A) $2a^2$ (B) $3a$ (C) $4a$ (D) $3a^2$ — 1

8 Using your calculator, find the value of $\sqrt[3]{60}$, correct to two decimal places.

(A) 3.91 (B) 7.74 (C) 7.75 (D) 20.00 — 1

9 What is the basic numeral for $5^3 \times 3^5$?

(A) 225 (B) 30 375 (C) 16 777 216 (D) 2 562 890 625

10 Which of these is written in ascending order?

(A) $\sqrt{50}, 6, \sqrt{43}, 7, \sqrt{31}$ (B) $\sqrt{43}, 7, \sqrt{50}, 6, \sqrt{31}$

(C) $6, \sqrt{31}, \sqrt{50}, 7, \sqrt{43}$ (D) $\sqrt{31}, 6, \sqrt{43}, 7, \sqrt{50}$ — 1

Total marks

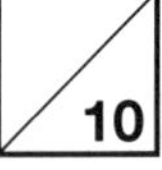

Indices

PRETEST

LEVELS 2 AND 3

Time allowed: 10 minutes **Total marks: 10**

Marks

1 What is the value of $3^2 \times 3^4$?

(A) 3^6 (B) 3^8 (C) 72 (D) 1088 — 1

2 Nick is asked to rewrite 72 as a product of its prime factors in index form. Which of these is the correct answer?

(A) $3^2 \times 2^2$ (B) $3^3 \times 2^3$ (C) $3^4 \times 2^3$ (D) $3^2 \times 2^3$ — 1

3 How many zeros are in the number $10^2 \times 10^3$?

(A) 5 (B) 6 (C) 7 (D) 8 — 1

4 Which of these is equal to $(3^2 \times 3)^3$?

(A) 3^5 (B) 3^6 (C) 3^8 (D) 3^9 — 1

5 Which of these is equal to $\frac{3.8971^4}{\sqrt{19 - 11.8}}$, correct to three significant figures?

(A) 8.59×10 (B) 8.60×10 (C) 8.59×10^2 (D) 8.60×10^2 — 1

6 Which of the following is the reciprocal of $\frac{4x^2}{y}$?

(A) $\frac{2x}{y}$ (B) $\frac{y}{4x^2}$ (C) $\frac{y}{2x}$ (D) $\frac{4y}{x^2}$ — 1

7 Which of these is equal to $(10^3)^2 \div (10^2)^2$?

(A) 10 (B) 10^2 (C) 10^4 (D) 10^6 — 1

8 What is the value of $2a^0 - (2a)^0$?

(A) -1 (B) 0 (C) 1 (D) 2 — 1

9 If $(8.6 \times 10^3) \times (4.8 \times 10^4) = 4.128 \times 10^x$, what is the value of x?

(A) 5 (B) 6 (C) 7 (D) 8 — 1

10 If the square of 2^a is 64, which of these is the value of a?

(A) 3 (B) 6 (C) 4 (D) 8 — 1

Total marks

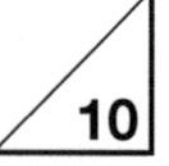

Indices

LEVEL 1 TEST — PART A

Time allowed: 40 minutes

Marks

1 Simplify $a^3 \times a^2$.

(A) a (B) a^5 (C) a^6 (D) $2a^5$ — 1

2 Simplify $y^6 \times y^2$.

(A) y^3 (B) y^4 (C) y^6 (D) y^8 — 1

3 Which of these is equal to $4m^2 \times 3m^4$?

(A) $12m^6$ (B) $7m^6$ (C) $7m^8$ (D) $12m^8$ — 1

4 What is the value of $10^5 \times 10^4$?

(A) 10^9 (B) 100^9 (C) 10^{20} (D) 100^{20} — 1

5 Which of these is the same as $\frac{p^6}{p^3}$?

(A) 2 (B) 3 (C) p^2 (D) p^3 — 1

6 Which of these has a value between 3 and 5?

(A) $\sqrt{4}$ (B) $\sqrt{12}$ (C) $\sqrt{26}$ (D) $\sqrt{32}$ — 1

7 Which of these is equal to $a \times a \times a \times b \times b$?

(A) a^3b^2 (B) $3a \times 2b$ (C) ab^5 (D) $5ab$ — 1

8 Simplify $(p^3)^2$.

(A) p^5 (B) $5p$ (C) $6p$ (D) p^6 — 1

9 Which of these is equal to $15a^5 \div 5a$?

(A) 3 (B) $3a^4$ (C) $10a^4$ (D) 10 — 1

10 What is the value of 4^0?

(A) 0 (B) 1 (C) 4 (D) 2 — 1

11 Which of the following is equal to $\left(\sqrt{4}\right)^2$?

(A) 1 (B) 2 (C) 4 (D) 16 — 1

12 What is the value of 3^6?

(A) 18 (B) 36 (C) 729 (D) 333 333 — 1

13 Li is asked to write $(2p^2q)^2$ in expanded form. Which of the following is Li's correct answer?

(A) $2 \times p \times p \times q$

(B) $4 \times p \times p \times q$

(C) $2 \times 2 \times p \times p \times p \times p \times q$

(D) $2 \times 2 \times p \times p \times p \times p \times q \times q$ — 1

LEVEL 1 TEST

PART B

Write answers and working in the space provided.

Marks

14 Simplify:

a $a^4 \times a^3$ b $4y^2 \times y$ c $6p^5 \times 3p^2$

3

15 Simplify:

a $x^5 \div x^2$ b $a^4 \div a^3$ c $5y^3 \div 5y$

3

16 Simplify:

a $(m^3)^2$ b $(p^3)^6$ c $(2w^3)^2$

3

17 Simplify:

a a^0 b $2p^0$ c $(b^2)^0$

3

18 Simplify:

a $g^4 \times g^2 \times g^6$ b $3a^2 \times 4a^6 \times 2a^3$ c $3y^2 \times 4y^6$

3

19 Which whole number is closest to:

a $\sqrt{15}$? b $\sqrt{32}$? c $2 \times \sqrt{5}$?

3

20 Write the following in index form:

a $2 \times 2 \times 2 \times 2$ b $p \times q \times p \times p \times q$ c $a \times a \times b \times b \times b \times ab \times ba$

3

21 Simplify:

a $-3a^2 \times 2a^4$ b $-6m^4 \div -2m$ c $(-p^2)^2$

3

22 Simplify:

a $m^2n^3 \times m^3$ b $x^3y^4 \div xy$ c $(3c^3d^2)^2$

3

Total marks 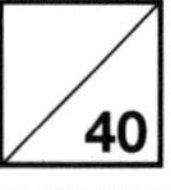 40

Indices

LEVEL 2 TEST — PART A

Time allowed: 40 minutes

Marks

1 Simplify $5x^3y \times 2x^2y^3$.

(A) $7x^5y^4$ (B) $7x^6y^3$ (C) $10x^5y^4$ (D) $10x^6y^3$ — 1

2 Expand $3a^2(2a^4 - b)$.

(A) $5a^6 - 3a^2b$ (B) $6a^6 - 3a^2b$ (C) $6a^8 - 3a^2b$ (D) $5a^8 - 3a^2b$ — 1

3 What is the value of $3a^0 - (2a)^0$?

(A) 1 (B) $3a - 1$ (C) 2 (D) 0 — 1

4 Simplify $\frac{15y^8}{3y^2}$.

(A) $12y^4$ (B) $12y^6$ (C) $5y^4$ (D) $5y^6$ — 1

5 Factorise $2p^5 - p^3$.

(A) p (B) p^2 (C) $p^3(2p^2 - 1)$ (D) $p^5(2p - 1)$ — 1

6 Which of these is equal to the expression $\frac{2}{m^3}$?

(A) $(2m)^{-3}$ (B) $2m^{-3}$ (C) $2m^{\frac{1}{3}}$ (D) $m^{\frac{2}{3}}$ — 1

7 Which of the following is equal to 0.003 71 written in scientific notation?

(A) 3.71×10^{-3} (B) 3.71×10^{-2} (C) 3.71×10^{-4} (D) 3.71×10^{-5} — 1

8 Simplify $12p^{-3} \div 4p^2$.

(A) $3p$ (B) $3p^{-1}$ (C) $3p^{-5}$ (D) $3p^5$ — 1

9 What is the value of $16^{-\frac{1}{2}}$?

(A) −8 (B) −4 (C) 0.25 (D) 4 — 1

10 What is the basic numeral for 8.03×10^{-3}?

(A) 0.803 (B) 0.0803 (C) 0.008 03 (D) 0.000 803 — 1

11 Which of these is **not** equal to 2?

(A) $2^5 \div 2^4$ (B) $4^{\frac{1}{2}}$ (C) $\left(\frac{1}{2}\right)^{-1}$ (D) $\frac{1}{2}(2^2 - 2)$ — 1

12 Simplify $(a^4)^3 \div (a^2)^4$.

(A) a^{-1} (B) a (C) a^2 (D) a^4 — 1

13 What is the value of $\left(\frac{81}{16}\right)^{-\frac{1}{4}}$?

(A) $\frac{2}{3}$ (B) $\frac{4}{9}$ (C) $1\frac{1}{2}$ (D) $2\frac{1}{4}$ — 1

Indices

LEVEL 2 TEST — PART B

Write answers and working in the space provided.

Marks

14 Simplify:

a $5m^4 \times 3m^2$ ______ b $(-6ab^2) \times 3a^2b^5$ ______ c $(-3x^5)^2$ ______ 3

15 Simplify:

a $12a^5 \div (-4a^2)$ ______ b $b^4 \div b^5$ ______ c $25p^3q \div 5pq$ ______ 3

16 Expand:

a $m^3(2m^5 + 4)$ ______ b $ab^3(ab^2 - 3b)$ ______ c $x^7y(y^3 - 2x^2)$ ______ 3

17 Simplify:

a $b^{-3} \times b^4$ ______ b $p^4 \div p^{-3}$ ______ c $(s^{-3})^2$ ______ 3

d $a^{\frac{3}{4}} \times a^{\frac{1}{4}}$ ______ e $m^{\frac{4}{5}} \div m^{\frac{3}{5}}$ ______ f $\left(p^{\frac{2}{3}}\right)^3$ ______ 3

18 Evaluate:

a 2^{-3} ______ b $\left(\frac{5}{8}\right)^{-1}$ ______ c $\left(\frac{2}{3}\right)^{-2}$ ______ 3

d $4^{\frac{1}{2}}$ ______ e $1000^{\frac{1}{3}}$ ______ f $\left(\frac{4}{25}\right)^{-\frac{1}{2}}$ ______ 3

19 Calculate, leaving your answer in scientific notation, correct to two significant figures:

a $\frac{1}{42 \times 38}$ ______ b $\sqrt{0.008}$ ______ c 0.1763^4 ______ 3

20 Use your calculator to find the value of x:

a $2^x = 1024$ ______ b $5^x = 15\ 625$ ______ c $1.6^x = 10.485\ 76$ ______ 3

Total marks

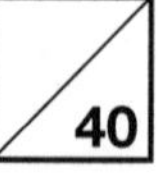

Indices

LEVEL 3 TEST PART A

Time allowed: 40 minutes

Marks

1 If 0.000 371 = 3.71×10^x, what is the value of x?

(A) 3 (B) 4 (C) −4 (D) −3 [1]

2 Simplify $b^{-3} \times b^4 \div b^{-2}$.

(A) b^{-1} (B) b^2 (C) b^3 (D) b^4 [1]

3 Simplify $2^{2a} \div 2^{-a}$.

(A) 1 (B) 2 (C) 2^a (D) 2^{3a} [1]

4 Which of these is equal to $2^{2n} - 2^n$?

(A) $2^n(2^n - 1)$ (B) 2^n (C) 2^2 (D) $2^n(n - 1)$ [1]

5 Which of these is equivalent to $\dfrac{3}{\sqrt{a-1}}$?

(A) $(a-1)^{\frac{3}{2}}$ (B) $(a-1)^{-\frac{3}{2}}$ (C) $3(a-1)^{\frac{1}{2}}$ (D) $3(a-1)^{-\frac{1}{2}}$ [1]

6 Which of these is equal to $\dfrac{\sqrt{12.89}}{(2.7 \times 13.9)^3}$, correct to three significant figures?

(A) 6.79×10^{-4} (B) 6.79×10^{-5} (C) 3.57×10^3 (D) 3.57×10^4 [1]

7 Simplify $(y^{2-a})^{-2} \times y^{-2a}$.

(A) $\dfrac{1}{y^4}$ (B) $\dfrac{4}{y^4}$ (C) $y^{4(a-1)}$ (D) $y^{4(a+1)}$ [1]

8 Which of these is a square root of $36x^{16}$?

(A) $6x^4$ (B) $6x^8$ (C) $18x^4$ (D) $18x^8$ [1]

9 Simplify 4×2^x.

(A) 8^x (B) 2^{x+1} (C) 2^{x+2} (D) 2^{x+3} [1]

10 Simplify $25^x \div 5^{-x}$.

(A) 5^{-3x} (B) 5^{-2x} (C) 5^{2x} (D) 5^{3x} [1]

11 Which of these is equal to $\sqrt{4.901 \times 10^{-3}}$, correct to three significant figures?

(A) 7×10^{-2} (B) 7×10^{-1} (C) 7.00×10^{-2} (D) 7.01×10^{-2} [1]

12 Simplify $\dfrac{4^m}{2^{3m}}$.

(A) $\dfrac{1}{2^m}$ (B) $\dfrac{2}{m}$ (C) 2^m (D) $\dfrac{m}{2}$ [1]

13 Simplify $\left(\dfrac{m^3}{8}\right)^{-\frac{2}{3}}$.

(A) $\dfrac{2}{m}$ (B) $\dfrac{4}{m^2}$ (C) $\dfrac{m}{2}$ (D) $\dfrac{m^2}{4}$ [1]

LEVEL 3 TEST

PART B

Write answers and working in the space provided.

Marks

14 Simplify:

a $\left(x^6y^9\right)^{\frac{2}{3}}$ **b** $\left(a^4b^{12}\right)^{\frac{3}{4}}$ **c** $\left(27p^6\right)^{-\frac{2}{3}}$

3

15 Simplify, expressing the answers with positive indices:

a $\left(9k^{-4}\right)^{\frac{1}{2}}$ **b** $\left(8y^3\right)^{-\frac{1}{3}}$ **c** $\left(\frac{m}{4}\right)^{-\frac{1}{2}}$

3

16 If $a = \frac{2}{3}$, $b = \frac{1}{2}$ and $c = \frac{1}{3}$, find the value of:

a $(abc)^{-2}$ **b** $\left(\frac{ab}{c}\right)^{-2}$ **c** $\frac{a^2b^3}{c^2}$

6

17 Find the value of the pronumeral:

a $2^{x+2} = 8$ **b** $3^{2a} = \frac{1}{3}$ **c** $5^{2y-1} = \sqrt{5}$

6

18 Simplify:

a $4^{2x} \div 8^x$ **b** $2^{x+1} \div 2^{x-2}$

6

19 Simplify $\left(\frac{36x^7y^3}{4x^3y^5}\right)^{-\frac{1}{2}}$.

3

Total marks 40

Linear Equations

PRETEST

LEVEL 1

Time allowed: 10 minutes **Total marks: 10**

Marks

1 Which of these is equal to $4p - 2 - p$?

(A) p (B) $3p - 2$ (C) $5p - 2$ (D) $5p^2 - 2$ 1

2 Expand $3(4m + 1)$.

(A) $7m + 3$ (B) $7m + 4$ (C) $12m + 3$ (D) $15m$ 1

3 What is the value of $2a + 1$ when $a = 4$?

(A) 7 (B) 9 (C) 10 (D) 25 1

4 If $b + 3 = 11$, what is the value of b?

(A) –8 (B) 6 (C) 8 (D) 14 1

5 Simplify $\frac{6w}{6}$.

(A) w (B) 6 (C) $5w$ (D) w^5 1

6 The expression $(5c - 1)$ is multiplied by 2. What is the result?

(A) $10c - 2$ (B) $9c$ (C) $12c - 1$ (D) $10c - 1$ 1

7 When the equation $t - 5 = 3$ is solved, what is the value of t?

(A) –2 (B) 2 (C) 8 (D) 15 1

8 When a secret number is added to 8 the result is 13. Which of these is the secret number?

(A) 3 (B) 5 (C) 6 (D) 7 1

9 Solve the equation $\frac{x}{3} = 6$.

(A) $x = 2$ (B) $x = 3$ (C) $x = 9$ (D) $x = 18$ 1

10 Which of these equations has the solution of $b = 3$?

(A) $b - 3 = 6$ (B) $\frac{b}{4} = 1$ (C) $b + 5 = -2$ (D) $5b = 15$ 1

Total marks /10

Linear Equations

PRETEST

LEVELS 2 AND 3

Time allowed: 10 minutes **Total marks: 10**

Marks

1 Expand and simplify $3y - (2y - 1)$.

(A) $y + 1$ (B) $y - 1$ (C) $y - 2$ (D) $y - 3$ [1]

2 Which of these is the solution to the equation $2m - 5 = 9$?

(A) $m = 2$ (B) $m = 4$ (C) $m = 7$ (D) $m = 14$ [1]

3 If $p = 3$ and $q = -2$, what is the value of $4p + 3q$?

(A) 6 (B) 8 (C) 10 (D) 11 [1]

4 If $x = 5$ and $2x - y = 12$, what is the value of y?

(A) –2 (B) 2 (C) 5 (D) 15 [1]

5 The area of a rectangle is 30 cm^2. If the dimensions of the rectangle are $(k + 2)$ cm and 5 cm, what is the value of k?

(A) $k = 3$ (B) $k = 4$ (C) $k = 5$ (D) $k = 6$ [1]

6 Which of these is the solution of $\frac{x}{-3} = -6$?

(A) $x = -18$ (B) $x = -2$ (C) $x = 2$ (D) $x = 18$ [1]

7 If x is an odd number, which of these expressions is the next consecutive odd number?

(A) y (B) z (C) $x + 1$ (D) $x + 2$ [1]

8 Which of the following equations has a solution of $p = 3$ and $q = 5$?

(A) $p + 2q = 8$ (B) $2p - q = 1$ (C) $p - 2q = 1$ (D) $3p - 2q = 4$ [1]

9 If $\frac{2}{y} = 4$, what is the value of y?

(A) $y = \frac{1}{8}$ (B) $y = \frac{1}{2}$ (C) $y = 2$ (D) $y = 8$ [1]

10 Christine checked the prices at the school canteen. She found that a bottle of water and a wrap costs \$5.30, while two bottles of water and a wrap costs \$7.80. What is the cost of a wrap?

(A) \$2.50 (B) \$2.60 (C) \$2.70 (D) \$2.80 [1]

Total marks 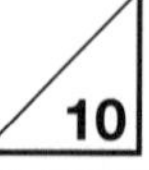 10

Linear Equations

LEVEL 1 TEST

PART A

Time allowed: 40 minutes

Marks

1 What is the solution of $x - 6 = -3$?

(A) $x = -9$ (B) $x = -3$ (C) $x = 3$ (D) $x = 9$ 1

2 Solve $2a + 6 = 10$.

(A) $a = 1$ (B) $a = 2$ (C) $a = 4$ (D) $a = 8$ 1

3 Which equation has a solution of $p = 4$?

(A) $5p = 9$ (B) $p + 6 = 2$ (C) $2p + 1 = 3$ (D) $p - 2 = 2$ 1

4 Solve the equation $\frac{y+1}{2} = 3$.

(A) $y = 1$ (B) $y = 5$ (C) $y = 6$ (D) $y = 7$ 1

5 What is the missing number?

$2 \times \square + 5 = 11$

(A) 1 (B) 2 (C) 3 (D) 6 1

6 By letting the number be x, write an equation to represent this statement: 'The product of a number and four is equal to eight'.

(A) $x + 4 = 8$ (B) $x - 4 = 8$ (C) $4x = 8$ (D) $\frac{x}{4} = 8$ 1

7 What is the value of m in the equation $2(m + 1) = 8$?

(A) $m = 3$ (B) $m = 4$ (C) $m = 5$ (D) $m = 7$ 1

8 For which of these equations is the solution $a = 0$?

(A) $\frac{a}{5} = 5$ (B) $5a = 5$ (C) $a - 5 = 5$ (D) $a + 5 = 5$ 1

9 Which of these equations does **not** have a solution of $p = 2$?

(A) $8p = 16$ (B) $2p + 2 = 6$ (C) $4 - p = 2$ (D) $\frac{p}{2} = 4$ 1

10 An adult movie ticket and a student movie ticket cost a total of \$31. If an adult ticket costs \$18, what is the cost of two student tickets?

(A) \$13 (B) \$15 (C) \$26 (D) \$30 1

11 Solve the equation $\frac{b}{2} = -5$.

(A) $b = -10$ (B) $b = -3$ (C) $b = 3$ (D) $b = 10$ 1

12 What is the solution of the equation $12 = 2x + 4$?

(A) $x = 4$ (B) $x = 6$ (C) $x = 8$ (D) $x = 10$ 1

13 Raelene solved an equation correctly. If her answer was $x = -2$, which of these could be her equation?

(A) $3x + 1 = 7$ (B) $4x = 8$ (C) $2 + x = 0$ (D) $2x + 4 = 8$ 1

Linear Equations

LEVEL 1 TEST — PART B

Write answers and working in the space provided.

Marks

14 Solve:

a $2a + 1 = 9$ b $3p - 2 = 13$ c $4m + 3 = 7$

15 Solve:

a $\frac{p-5}{2} = 4$ b $\frac{t+3}{4} = 5$ c $\frac{d+5}{3} = -2$

16 Solve:

a $2b - 4 = b + 3$ b $3m + 2 = m + 6$ c $5n + 1 = 2n - 5$

17 Solve:

a $2(a + 1) = 16$ b $3(2p - 3) = 9$ c $5(3k + 1) = 20$

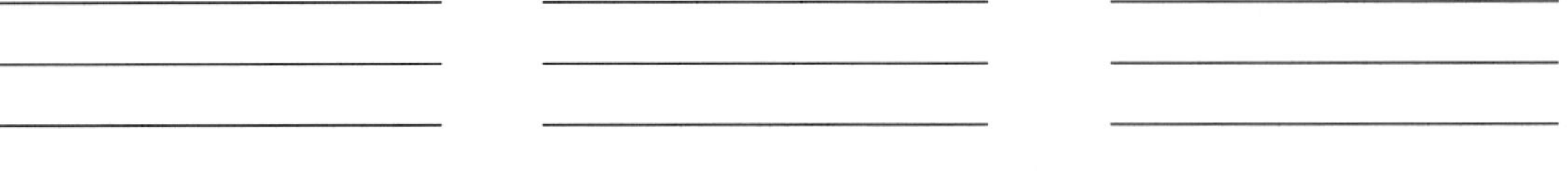

18 Solve $\frac{2z-1}{3} = 5$.

3

Total marks

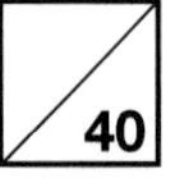

Linear Equations

LEVEL 2 TEST **PART A**

Time allowed: 40 minutes

Marks

1 What is the solution of $-2(m-4) = -4$?

(A) $m = 0$ (B) $m = 2$ (C) $m = 4$ (D) $m = 6$ 1

2 Which of these equations has a solution of $x = -1$?

(A) $\frac{2x-1}{3} = -1$ (B) $2(x+3) = -1$ (C) $4 = 2x + 5$ (D) $\frac{3x-1}{2} = 4$ 1

3 What is the value of y when $x = 2$ is substituted into $2x + y = 9$?

(A) 1 (B) 5 (C) 7 (D) 8 1

4 Solve the simultaneous equations $p + q = 4$ and $p - q = 2$.

(A) $p = 3, q = 1$ (B) $p = 1, q = 1$ (C) $p = 1, q = 3$ (D) $p = 2, q = 2$ 1

5 Which of these is the solution of the equation $3 - (k + 1) = 5$?

(A) $k = -3$ (B) $k = -2$ (C) $k = -1$ (D) $k = 1$ 1

6 Solve the equation $\frac{3x-1}{2} = x$.

(A) $x = -1$ (B) $x = 1$ (C) $x = 2$ (D) $x = 3$ 1

7 If $p = 3$ is substituted into $2p - q = 8$, what is the value of q?

(A) $q = -4$ (B) $q = -2$ (C) $q = 2$ (D) $q = 4$ 1

8 Which of the following equations is equivalent to $\frac{2x-5}{3} = 6$?

(A) $2x - 15 = 6$ (B) $2x - 5 = 18$ (C) $2x - 15 = 2$ (D) $\frac{2x}{3} = 11$ 1

9 Which of these is the solution of the equation $3(d-1) = 2(d+2)$?

(A) $d = 3$ (B) $d = 5$ (C) $d = 12$ (D) $d = 7$ 1

10 Solve the equation $\frac{2y}{3} + 3 = 6$.

(A) $y = \frac{1}{2}$ (B) $y = 4\frac{1}{2}$ (C) $y = 6$ (D) $y = 13\frac{1}{2}$ 1

11 Substitute $s = -3$ in $t = 2 - 4s$ to find the value of t.

(A) $t = -14$ (B) $t = -10$ (C) $t = 10$ (D) $t = 14$ 1

12 Solve the equation $\frac{-1}{q} = 2$.

(A) $q = -2$ (B) $q = -\frac{1}{2}$ (C) $q = \frac{1}{2}$ (D) $q = 2$ 1

13 Which of these equations has a solution of $y = 1$?

(A) $\frac{y+1}{3} = 1$ (B) $\frac{y-1}{3} = 1$ (C) $\frac{2y+1}{3} = 1$ (D) $\frac{2y-1}{3} = 1$ 1

Linear Equations

LEVEL 2 TEST — PART B

Write answers and working in the space provided.

Marks

14 Solve:

a $3t + 1 = t - 5$

b $6a - 4 = 2a + 6$

c $2m + 3 = 8m - 3$

6

15 Solve:

a $2(2q - 5) - 3(4q + 1) = 3$

b $12p - (2p + 2) = 4p$

c $3(4 - 2y) - 9 = -3(4y + 1)$

9

16 Solve:

a $\frac{5x+1}{2} = 4x - 7$

b $\frac{4p+2}{3} - \frac{3p}{2} = 3$

c $\frac{1-a}{4} - \frac{3+a}{3} = 1$

9

17 Solve simultaneously $3x + 2y = 7$ and $3x - y = 1$.

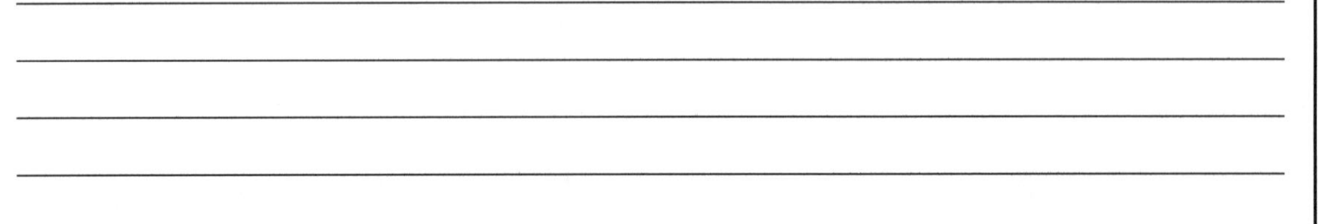

3

Total marks

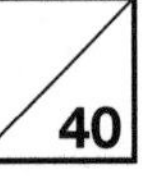

40

Linear Equations

LEVEL 3 TEST — PART A

Time allowed: 40 minutes

Marks

1 Consider the equation $\frac{2x}{3} - 4 = \frac{5x}{2} + 1$. Which of the following would be a correct step in solving this equation?

(A) $\frac{2x}{3} - 3 = \frac{5x}{2}$ (B) $\frac{2x}{3} = \frac{5x}{2} + 5$ (C) $2x - 4 = \frac{15x}{2} + 3$ (D) $\frac{4x}{6} - 8 = 5x + 2$ — 1

2 Tom has two sisters. He is 4 years older than Gemma and 6 years older than Steph. If the sum of their three ages is 47, how old is Gemma?

(A) 13 (B) 14 (C) 15 (D) 19 — 1

3 Which of these pairs of equations has solutions $a = 2$ and $b = 3$?

(A) $a + b = 5$ and $a - b = 1$ (B) $2a - b = 5$ and $a - 3b = -7$

(C) $3a + b = 9$ and $a - b = -1$ (D) $a - b = -1$ and $a + 2b = 5$ — 1

4 Solve the equation $\frac{a}{3} + \frac{a}{2} = 10$.

(A) $a = \frac{5}{6}$ (B) $a = 6$ (C) $a = 12$ (D) $a = 24$ — 1

5 If $4p = 20$ and $q = 8 - 3p$, what is the value of $2q$?

(A) -104 (B) -14 (C) -7 (D) -4 — 1

6 Which of these is the solution of $\frac{1}{k+1} = 2$?

(A) $k = -1.5$ (B) $k = -0.5$ (C) $k = 0.5$ (D) $k = 1.5$ — 1

7 Solve the simultaneous equations $y = 2x$ and $y = x - 1$.

(A) $x = 1, y = 2$ (B) $x = 2, y = 4$ (C) $x = -2, y = -4$ (D) $x = -1, y = -2$ — 1

8 Two less than three times a number is equal to four more than twice the number. What is the number?

(A) 1 (B) 2 (C) 4 (D) 6 — 1

9 What is the value of a if $\frac{3a}{4} = \frac{2}{3}$?

(A) $a = \frac{8}{9}$ (B) $a = \frac{3}{8}$ (C) $a = \frac{1}{2}$ (D) $a = 1\frac{1}{8}$ — 1

10 If $2 - c = 7$ and $d = 2c + 1$, what is the value of d?

(A) $d = -11$ (B) $d = -9$ (C) $d = 11$ (D) $d = 15$ — 1

11 If $m = n + 1$ and $m + n = 5$, which of these is the value of n?

(A) $n = 2$ (B) $n = 3$ (C) $n = 4$ (D) $n = 6$ — 1

12 Solve $\sqrt{2x + 5} = 3$.

(A) $x = 1$ (B) $x = 2$ (C) $x = 3$ (D) $x = 4$ — 1

13 Half of Lisa's age 4 years ago equals her age 12 years ago. How old is Lisa now?

(A) 20 (B) 24 (C) 28 (D) 32 — 1

Linear Equations

LEVEL 3 TEST — PART B

Write answers and working in the space provided.

Marks

14 Solve the following equations:

a $\frac{2a+1}{3} - 4 = 11 - 3a$ **b** $\frac{3k-1}{4} - \frac{2k+5}{3} = 2k$ **c** $\frac{p+1}{2} - \frac{3-2p}{3} = p - 1$

9

15 Solve the following equations:

a $\frac{2y-4}{y} - 1 = 2$ **b** $\frac{3}{4b+6} = \frac{1}{10}$ **c** $\frac{4a-1}{2a+1} = 1$

6

16 Solve these simultaneous equations:

a $y = 2x + 1$ and $x - 2y - 4 = 0$ **b** $5a + 2b = -4$ and $3a + b = -3$ **c** $2m - 3n = -3$ and $3m + 2n = -11$

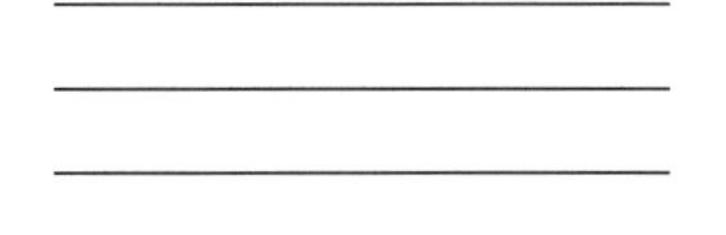

9

17 The ticket price for a concert is \$89 for adults and \$69 for children. A total of \$167 000 was collected from the 2000 tickets sold. Use simultaneous equations to find the number of adults who attended the concert.

3

Total marks

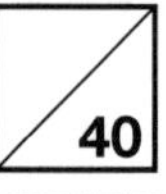

40

Inequalities and Formulae

PRETEST — LEVEL 1

Time allowed: 10 minutes — **Total marks: 10**

Marks

1 Which of these numbers is between –4 and 2?

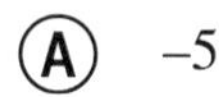

Ⓐ –5 Ⓑ –2 Ⓒ 2 Ⓓ 4 — 1

2 Which of these shows the solution of $x - 3 = 0$?

Ⓐ 

Ⓑ 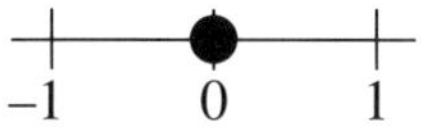

Ⓒ 

Ⓓ 

— 1

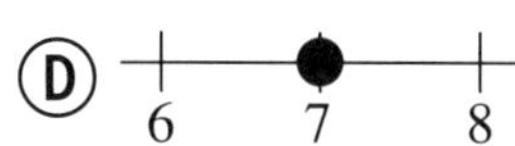

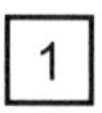

3 Which of these is a number greater than –3?

Ⓐ –8 Ⓑ –4 Ⓒ –3 Ⓓ –1 — 1

4 Which of these is the smallest?

Ⓐ –3 Ⓑ 1 Ⓒ 4 Ⓓ 12 — 1

5 Which of these is a solution of the equation $2a + 1 = 5$?

Ⓐ $a = 1$ Ⓑ $a = 2$ Ⓒ $a = 4$ Ⓓ $a = 6$ — 1

6 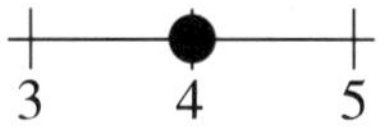

The graph shows the solution of an equation. Which of these could be the equation?

Ⓐ $6 - x = 2$ Ⓑ $2x = 4$ Ⓒ $x + 4 = 4$ Ⓓ $x - 4 = 4$ — 1

7 If $a = 3$ and $b = 5$ what is the value of $2ab$?

Ⓐ 10 Ⓑ 30 Ⓒ 60 Ⓓ 235 — 1

8 The formula for the area A of a square with side s is $A = s^2$. What is the value of A when $s = 8$?

Ⓐ $A = 4$ Ⓑ $A = 16$ Ⓒ $A = 64$ Ⓓ $A = 88$ — 1

9 Which of these inequalities is correct?

Ⓐ $-2 \leq -3$ Ⓑ $4 \leq -5$ Ⓒ $-4 \geq 2$ Ⓓ $-5 \leq 3$ — 1

10 The area (A) of a trapezium is found using the formula $A = \frac{h}{2}(a + b)$.

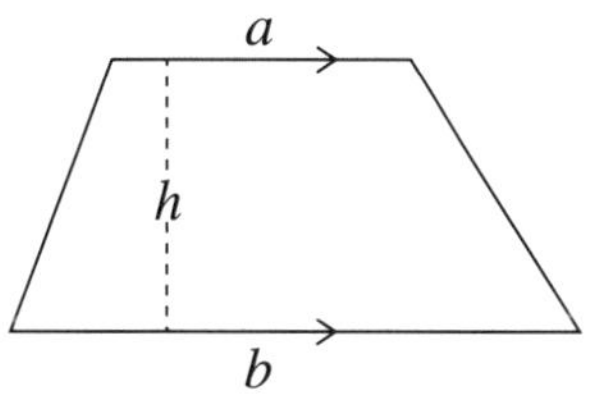

Which of these is used to find the area of a trapezium where $a = 8$, $b = 10$ and $h = 6$?

Ⓐ $\frac{8}{2}(10 + 6)$ Ⓑ $\frac{8}{2}(6 + 10)$ Ⓒ $\frac{6}{2}(8 + 10)$ Ⓓ $\frac{10}{2}(6 + 8)$ — 1

Total marks ___ / 10

Inequalities and Formulae

PRETEST — LEVELS 2 AND 3

Time allowed: 10 minutes — **Total marks: 10**

Marks

1 If $s = ut + \frac{1}{2}at^2$, which of these is the value of s when $u = 6$, $a = 10$ and $t = 4$?

(A) 104 (B) 144 (C) 224 (D) 824 — 1

2 Which of these shows the solution of $2(p - 5) = p + 1$?

(A) $p = -9$ (B) $p = -4$ (C) $p = 9$ (D) $p = 11$ — 1

3 The area A of a triangle is given as $A = \frac{1}{2}bh$, where b = base and h = height. What is the height of the triangle when the area is 48 cm² and the base is 8 cm?

(A) 3 cm (B) 6 cm (C) 8 cm (D) 12 cm — 1

4 Which of these is the solution of $\frac{2x + 7}{3} = 1$?

(A) $x = -3$ (B) $x = -2$ (C) $x = 5$ (D) $x = 10$ — 1

5 Which of these is a rearrangement of the equation $P = \frac{m - n}{2}$?

(A) $m = 2P + n$ (B) $m = 2P - n$ (C) $m = n - 2P$ (D) $m = \frac{P - n}{2}$ — 1

6 Which of these shows the solution of the equation $3(x - 1) = 2(1 - x)$?

(A) number line 0, 1, 2 with dot at 1 (B) number line −2, −1, 0 with dot at −1 (C) number line −4, −3, −2 with dot at −3 (D) number line 1, 2, 3 with dot at 2 — 1

7 The formula $S = \frac{D}{T}$ is used to find the speed S for a given distance D and time T.

Use the formula to find the distance travelled by a car travelling at 80 km/h for 4 hours.

(A) 20 km (B) 240 km (C) 280 km (D) 320 km — 1

8 Which of these equations has a solution of $w = 3$?

(A) $2w + 3 = 3$ (B) $3(w - 3) = 6$ (C) $\frac{1 - w}{3} = 1$ (D) $\frac{3}{w} = 1$ — 1

9 If $d = 0.2v + 0.004v^2$, which of these is the value of d when $v = 0.8$?

(A) $d = 0.160\,010\,24$ (B) $d = 0.130\,56$ (C) $d = 0.162\,56$ (D) $d = 0.202\,56$ — 1

10 If $m = 12$ and $v = 4$, what is the value of E when $E = \frac{1}{2}mv^2$?

(A) $E = 24$ (B) $E = 96$ (C) $E = 576$ (D) $E = 1152$ — 1

Total marks 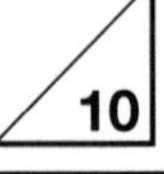

Inequalities and Formulae

LEVEL 1 TEST — PART A

Time allowed: 40 minutes

Marks

1 Solve $a - 4 > 6$.

(A) $a < 2$ (B) $a > 2$ (C) $a < 10$ (D) $a > 10$ — 1

2 Which of these is the solution of $4m \le -8$?

(A) $m \le -4$ (B) $m \le -2$ (C) $m \ge -2$ (D) $m \ge -4$ — 1

3 If $d = \frac{m}{v}$, what is the value of d when $m = 10$ and $v = 4$?

(A) $d = 0.4$ (B) $d = 2.5$ (C) $d = 6$ (D) $d = 40$ — 1

4 Solve the equation $\frac{x+2}{3} < 1$.

(A) $x < 1$ (B) $x > 1$ (C) $x < 5$ (D) $x > 5$ — 

5 Which of these represents the solution of $p + 3 \ge 2$?

(A) –2 –1 0 (B) –2 –1 0 (C) 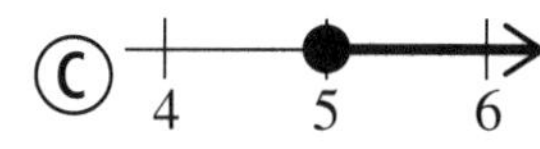

(D) 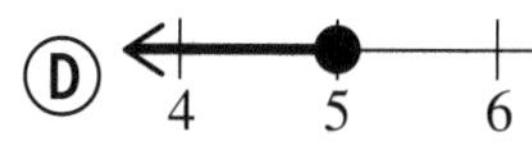

— 1

6 If $D = ST$, what is the value of D when $S = 80$ and $T = 0.5$?

(A) $D = 8$ (B) $D = 40$ (C) $D = 85$ (D) $D = 400$ — 1

7 –4 –3 –2

Which of these inequalities is represented by the graph above?

(A) $x \le -3$ (B) $x < -3$ (C) $x \ge -3$ (D) $x > -3$ — 1

8 Solve the inequality $3 + y \ge 5$.

(A) $y \ge 2$ (B) $y \le 2$ (C) $y \le 8$ (D) $y \ge 8$ — 1

9 Which of these has P as the subject of a formula?

(A) $PQ = M + N$ (B) $T = 5P - 1$ (C) $P^2 = 25AB$ (D) $P = C(D + T)$ — 1

10 Use the formula $C = 2\pi r$ to find C, correct to two decimal places when $r = 10$.

(A) $C = 15.14$ (B) $C = 62.83$ (C) $C = 125.66$ (D) $C = 314.16$ — 1

11 Solve the inequality $2p - 1 > 5$.

(A) $p < 2$ (B) $p > 2$ (C) $p < 3$ (D) $p > 3$ — 1

12 If $t = \frac{v-u}{a}$, what is the value of t when $v = 80$, $u = 60$ and $a = 10$?

(A) $t = 2$ (B) $t = 10$ (C) $t = 12$ (D) $t = 74$ — 1

13 If $A = LB$, what is the value of B if $A = 120$ and $L = 12$?

(A) $B = 10$ (B) $B = 108$ (C) $B = 132$ (D) $B = 1440$ — 1

Inequalities and Formulae

LEVEL 1 TEST — PART B

Write answers and working in the space provided.

Marks

14 On separate number lines graph the following inequalities:

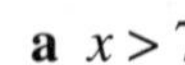

a $x > 7$ **b** $x \leq 3$ **c** $x \geq -3$

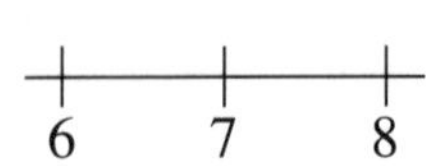

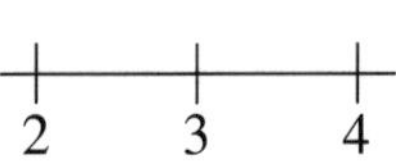

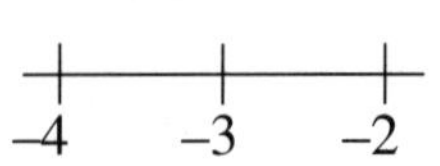

3

15 Write the inequality in terms of x that matches the graph:

a

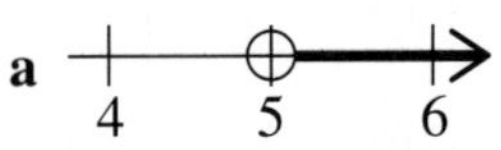

b

c –2 –1 0

3

16 Solve the inequalities:

a $3m + 2 > 8$ **b** $2p - 5 \geq 1$ **c** $\frac{a+3}{4} < 2$

6

17 Solve the inequalities:

a $-3y < 6$ **b** $-8a \geq 24$ **c** $\frac{q}{-4} < 6$

6

18 If $v = u + at$, what is the value of v when:

a $u = 6, a = 10, t = 3$? **b** $u = 5.2, a = 9.8, t = 4.1$? **c** $u = \frac{3}{5}, a = 9\frac{4}{5}, t = \frac{5}{7}$?

6

19 The formula $F = 32 + \frac{9}{5}C$ is used to convert temperatures in degrees, from Celsius (C) to Fahrenheit (F). What is the temperature in Fahrenheit when it is 30 °C?

3

Total marks

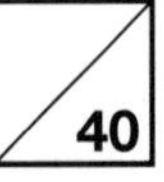

Inequalities and Formulae

LEVEL 2 TEST — PART A

Time allowed: 40 minutes

Marks

1 If $S = \frac{D}{T}$, what is the value of D when $S = 20$ and $T = 4$?

(A) $D = 0.2$ (B) $D = 5$ (C) $D = 24$ (D) $D = 80$ — 1

2 Which of these is the solution of $2b + 6 \leq b - 3$?

(A) $b \leq -9$ (B) $b \geq -9$ (C) $b \leq -3$ (D) $b \geq -3$ — 1

3 Which of these has the solution $m > -1$?

(A) $4m < -4$ (B) $-4m < -4$ (C) $-4m < 4$ (D) $\frac{m}{4} > -4$ — 1

4 Which of the following is the graph of $-2 < x \leq 1$?

(A) number line −2 to 1: open circle at −2, open circle at 1
(B)  number line −2 to 1: closed circle at −2, open circle at 1
(C) number line −2 to 1: open circle at −2, closed circle at 1
(D)  number line −2 to 1: closed circle at −2, closed circle at 1 — 1

5 Which of the following correctly expresses r (where $r > 0$) as the subject of $A = \pi r^2$?

(A) $r = \frac{A}{\pi}$ (B) $r = \sqrt{\frac{A}{\pi}}$ (C) $r = \frac{\sqrt{A}}{\pi}$ (D) $r = \frac{A}{\sqrt{\pi}}$ — 1

6 If $m = \frac{a+b}{2}$, what is the value of b when $m = 8$ and $a = -4$?

(A) $b = 2$ (B) $b = 8$ (C) $b = 12$ (D) $b = 20$ — 1

7 Solve the inequality $3a + 3 \geq a - 5$.

(A) $a \geq -4$ (B) $a \geq -2$ (C) $a \geq -1$ (D) $a \geq 1$ — 1

8 Which of these is the solution of the inequality $6 - y > 4$?

(A)  number line −3, −2, −1: open circle at −2, arrow to the left
(B)  number line −3, −2, −1: open circle at −2, arrow to the right
(C)  number line 1, 2, 3: open circle at 2, arrow to the left
(D)  number line 1, 2, 3: open circle at 2, arrow to the right — 1

9 Which of the following correctly expresses b as the subject of $P = 2(l + b)$?

(A) $b = \frac{P - 2l}{2}$ (B) $b = \frac{2l - P}{2}$ (C) $b = \frac{2P - l}{2}$ (D) $b = \frac{l - 2P}{2}$ — 1

10 Use the formula $V = \frac{1}{3}Ah$, to find the value of h if $V = 24$ and $A = 18$.

(A) $h = 4$ (B) $h = 12$ (C) $h = 36$ (D) $h = 60$ — 1

11 Solve the inequality $4(2m - 1) \geq 12$.

(A) $m \geq 2$ (B) $m \geq 3$ (C) $m \geq 4$ (D) $m \geq 23$ — 1

12 If $c = \sqrt{\frac{a}{b}}$, what is the value of a when $b = 4$ and $c = 2$?

(A) $a = 2$ (B) $a = 4$ (C) $a = 8$ (D) $a = 16$ — 1

13 Solve the inequality $\frac{t}{4} - 3 \leq 5$.

(A) $t \leq 2$ (B) $t \leq 8$ (C) $t \leq 32$ (D) $t \leq 64$ — 1

Inequalities and Formulae

LEVEL 2 TEST — PART B

Write answers and working in the space provided.

Marks

14 Solve the inequalities:

a $4a - 1 < 2a - 5$ b $3c + 4 \geq c + 2$ c $5p - 2 \geq 7p + 8$

6

15 Solve the inequalities:

a $3(b + 2) > 2(b - 3)$ b $5(m - 1) > 2(2m + 4)$ c $2 + 3(2y - 1) \geq 17$

6

16 Solve the inequalities and graph your solution on a number line:

a $\frac{2p - 3}{4} \leq 1$ b $\frac{a}{5} + a > 6$ c $\frac{8 - q}{3} \geq q$

9

17 If $M = t + ab$, find the value of:

a t, if $M = 28$, $a = 4$ and $b = 3$ b a, if $M = 18$, $t = 6$ and $b = 3$

4

18 Make y the subject in $2x - 3y = 13$.

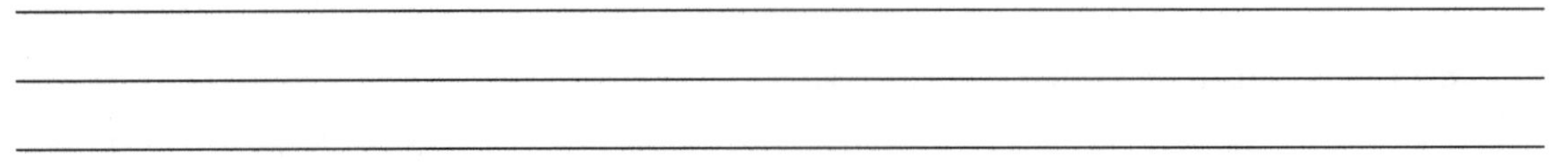

2

Total marks

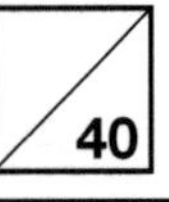

Inequalities and Formulae

LEVEL 3 TEST — PART A

Time allowed: 40 minutes

Marks

1 Which of the following correctly expresses b as the subject of $d = \frac{\sqrt{ab}}{c}$?

(A) $b = \frac{cd^2}{a}$ (B) $b = \frac{(cd)^2}{a}$ (C) $b = \frac{ad^2}{c}$ (D) $b = \frac{(ad)^2}{c}$ — 1

2 If $S = 2\pi r^2 + 2\pi rh$, find the value of h when $S = 992$, $r = 10$ and $\pi = 3.1$.

(A) $h = 3$ (B) $h = 4$ (C) $h = 6$ (D) $h = 8$ — 1

3 Which of the following correctly expresses q as the subject of $M = p\sqrt{2q}$?

(A) $q = \frac{M}{2p}$ (B) $q = \frac{M^2}{2p^2}$ (C) $q = \frac{\sqrt{M}}{2p^2}$ (D) $q = \sqrt{\frac{M}{2p^2}}$ — 1

4 Which equation correctly shows r as the subject of $S = \frac{40}{1-r}$?

(A) $r = \frac{40}{1-S}$ (B) $r = \frac{40}{S-1}$ (C) $r = \frac{S-40}{S}$ (D) $r = \frac{40-S}{S}$ — 1

5 Which of the following correctly expresses a as the subject of $s = ut + \frac{1}{2}at^2$?

(A) $a = \frac{2s - ut}{t^2}$ (B) $a = \frac{ut - 2s}{t^2}$ (C) $a = \frac{2(s - ut)}{t^2}$ (D) $a = \frac{2(ut - s)}{t^2}$ — 1

6 If $T = \frac{m^2 + 2n}{p}$, what is the value of p when $T = 12$, $m = 2$ and $n = 4$?

(A) $p = 1$ (B) $p = 2$ (C) $p = 8$ (D) $p = 12$ — 1

7 Which of these is the solution of the inequality $2 - 2x > -1 + x$?

(A) 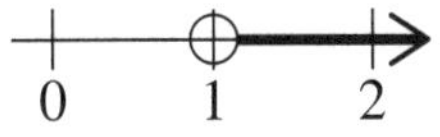

(B) 

(C) 

(D) 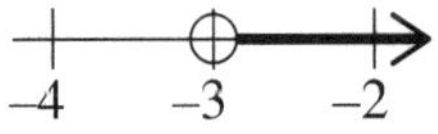

— 1

8 If $V = \pi r^2 h$, which of these is the value of h when $V = 192$ and $r = 16$?

(A) $h = \frac{3}{4\pi}$ (B) $h = \frac{4}{3\pi}$ (C) $h = \frac{3\pi}{4}$ (D) $h = \frac{4\pi}{3}$ — 1

9 Which of the following correctly expresses c as the subject of $E = mc^2$ when $c > 0$?

(A) $c = \sqrt{\frac{E}{m}}$ (B) $c = \frac{E}{\sqrt{m}}$ (C) $c = \frac{\sqrt{E}}{m}$ (D) $c = \sqrt{\frac{m}{E}}$ — 1

10 If $h = ut - 4.9t^2$, what is the value of u if $h = 12$ and $t = 4$?

(A) $u = 18.56$ (B) $u = 19.94$ (C) $u = 22.6$ (D) $u = 99.04$ — 1

11 Using $A = 6s^2$ and $V = s^3$, what is the value of V when $A = 54$ and $s > 0$?

(A) $V = 3$ (B) $V = 9$ (C) $V = 18$ (D) $V = 27$ — 1

12 If $S = \frac{n}{2}(a + l)$, what is the value of a if $S = 96$, $n = 48$ and $l = 18$?

(A) $a = -22$ (B) $a = -14$ (C) $a = 4$ (D) $a = 8$ — 1

13 Which of these is **not** a rearrangement of the formula $m = \frac{y-a}{x-b}$?

(A) $x = b + \frac{y-a}{m}$ (B) $a = y - m(x - b)$ (C) $y = m(x - b) + a$ (D) $b = x + \frac{y-a}{m}$ — 1

Inequalities and Formulae

LEVEL 3 TEST

PART B

Write answers and working in the space provided.

Marks

14 Solve the following inequalities:

a $\frac{3-2x}{2} - \frac{1-3x}{5} > 1$ **b** $\frac{4-a}{3} \geq \frac{3a+1}{2}$ **c** $\frac{3m}{4} - m < 2(1-m)$

9

15 Express y as the subject of the following:

a $z = \pi y + 2y - 2x$ **b** $z = \frac{2xy}{x-y}$ **c** $\frac{1}{x} + \frac{1}{y} = \frac{1}{z}$

9

16 a Given the formula $F = \frac{M(v^2 - u^2)}{2s}$, find the value of s if $F = 12.8$, $v = 12$, $u = 8$ and $M = 3.2$.

3

b Given the formula $S = x^2 + x\sqrt{4h^2 + x^2}$, find the value of positive h, if $S = 96$ and $x = 6$.

3

c Given the formula $A = \sqrt{s(s-a)(s-b)(s-c)}$, find the value of c if $A = 24$, $s = 16$, $a = 4$ and $b = 13$.

3

Total marks

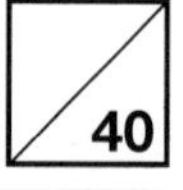

40

Linear Relationships

PRETEST

LEVEL 1

Time allowed: 10 minutes **Total marks: 10**

Marks

1 Which of these is the coordinates of the origin? 1

(A) (1, 1) (B) (0, 5) (C) (0, 0) (D) (5, 5)

2 If $x = 3$, which of these is the value of y in the equation $y = 2x + 1$? 1

(A) 5 (B) 6 (C) 7 (D) 24

The number plane is used to answer questions 3 to 10.

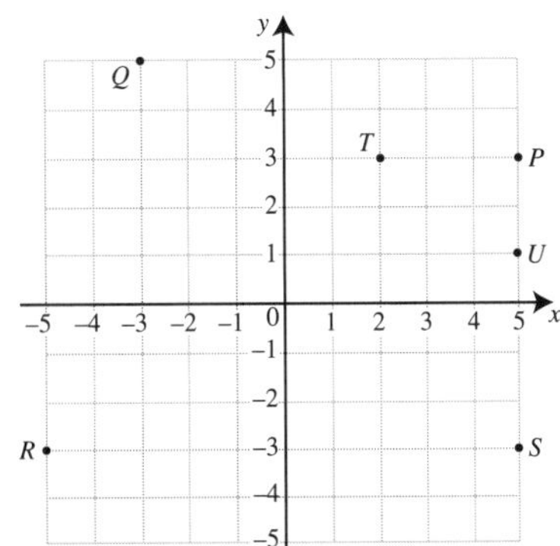

3 Which point is located at (–3, 5)? 1

(A) P (B) Q (C) R (D) S

4 How many units are there between P and T on the number plane? 1

(A) 2 (B) 3 (C) 4 (D) 5

5 What are the coordinates of the point R? 1

(A) (–3, 5) (B) (–5, 3) (C) (–3, –5) (D) (–5, –3)

6 How many units are there between R and S on the number plane? 1

(A) 5 (B) 9 (C) 10 (D) 11

7 A line passes through the points P and U. Which other point is on the line? 1

(A) R (B) T (C) Q (D) S

8 Which of these is the middle of the interval RS? 1

(A) (0, –3) (B) (–3, 0) (C) (0, 3) (D) (3, 3)

9 If the points R, S and U are joined, what is the name of RSU? 1

(A) equilateral triangle (B) right-angled triangle (C) isosceles triangle (D) rectangle

10 If T, P and U are three vertices of a rectangle, what are the coordinates of the fourth vertex of the rectangle? 1

(A) (2, 1) (B) (1, 2) (C) (–1, 2) (D) (1, –2)

Total marks

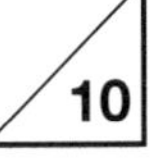

Linear Relationships

PRETEST

LEVELS 2 AND 3

Time allowed: 10 minutes **Total marks: 10**

Marks

1 Which of these is the middle of –4 and 6?

(A) –2 (B) –1 (C) 1 (D) 2 — 1

2 If $y = 3x - 1$ and $x = 2$, what is the value of y?

(A) 1 (B) 2 (C) 3 (D) 5 — 1

3 Which of these points is on the line with equation $4x - 3y + 1 = 0$?

(A) (3, 5) (B) (2, –3) (C) (2, 5) (D) (–1, –1) — 1

The number plane contains the lines p, q and r and is used to answer questions 4 to 10.

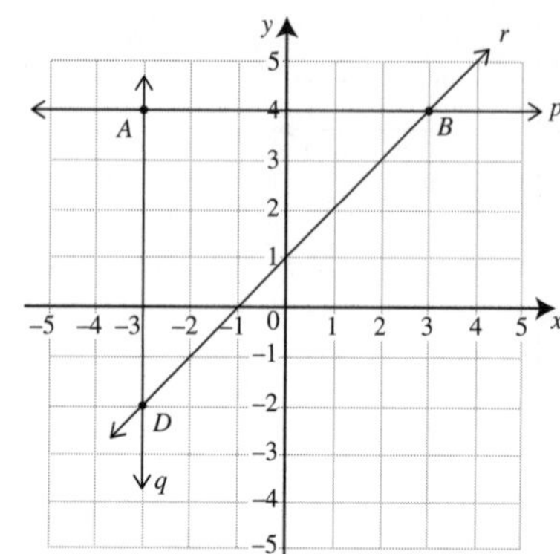

4 Which of these is the equation of the line p?

(A) $x = -3$ (B) $y = -3$ (C) $x = 4$ (D) $y = 4$ — 1

5 Which of these is the equation of the line q?

(A) $x = -3$ (B) $y = -3$ (C) $x = 4$ (D) $y = 4$ — 1

6 What is the equation of the line passing through (3, –2) and parallel to line p?

(A) $y = -2$ (B) $y = 3$ (C) $x = 3$ (D) $x = -2$ — 1

7 Which of these is the length of AB?

(A) 3 units (B) 4 units (C) 6 units (D) 7 units — 1

8 Which of these is the length of AD?

(A) 2 units (B) 6 units (C) 7 units (D) 8 units — 1

9 Which of these is the area of the triangle ABD?

(A) 12 units^2 (B) 18 units^2 (C) 24 units^2 (D) 36 units^2 — 1

10 At what point does the line p intersect with the line $x = 5$?

(A) (5, –3) (B) (–3, 5) (C) (5, 4) (D) (4, 5) — 1

Total marks 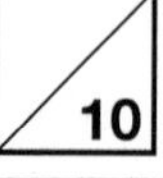 /10

Linear Relationships

LEVEL 1 TEST

PART A

Time allowed: 40 minutes

Marks

The number plane contains the lines a, b, c and d and is used to answer questions 1 to 13.

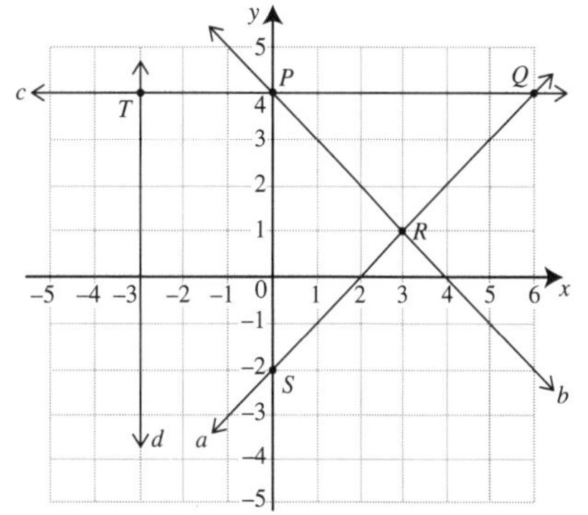

1 Which of these is the point of intersection of line a and line b? 1

Ⓐ (1, 3) Ⓑ (3, 1) Ⓒ (6, 4) Ⓓ (0, 4)

2 What is the y-intercept of the line a? 1

Ⓐ –2 Ⓑ 0 Ⓒ 2 Ⓓ 4

3 Which of these is the point of intersection of line b and line c? 1

Ⓐ (1, 3) Ⓑ (3, 1) Ⓒ (6, 4) Ⓓ (0, 4)

4 What is the equation of the line c? 1

Ⓐ $y = x + 4$ Ⓑ $x + y = 4$ Ⓒ $x = 4$ Ⓓ $y = 4$

5 What is the x-intercept of the line a? 1

Ⓐ –2 Ⓑ 0 Ⓒ 2 Ⓓ 4

6 Which of these is the gradient of line a? 1

Ⓐ –2 Ⓑ –1 Ⓒ 1 Ⓓ 2

7 What is the length of PS? 1

Ⓐ 2 units Ⓑ 4 units Ⓒ 6 units Ⓓ 7 units

8 What are the coordinates of the midpoint of PQ? 1

Ⓐ (3, 4) Ⓑ (0, 3) Ⓒ (3, 0) Ⓓ (4, 3)

9 What is the gradient of the line b? 1

Ⓐ –4 Ⓑ –1 Ⓒ 1 Ⓓ 4

10 Which of these is the gradient of line c? 1

Ⓐ 0 Ⓑ 1 Ⓒ 4 Ⓓ –4

11 Which of these points is on the line b? 1

Ⓐ (6, 0) Ⓑ (6, –1) Ⓒ (6, –2) Ⓓ (6, –3)

12 What is the equation of the line d? 1

Ⓐ $y = x - 3$ Ⓑ $x - y = -3$ Ⓒ $x = -3$ Ⓓ $y = -3$

13 What is the length of TQ? 1

Ⓐ 3 units Ⓑ 5 units Ⓒ 6 units Ⓓ 9 units

Linear Relationships

LEVEL 1 TEST — PART B

Write answers and working in the space provided.

Marks

14 Complete each table and graph the lines on separate number planes: 9

a $y = x + 3$

x	0	1	2
y			

b $y = 4 - x$

x	0	1	2
y			

c $y = 2x + 1$

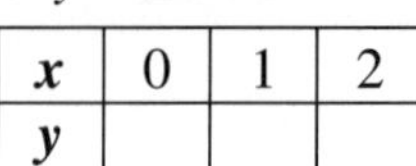

x	0	1	2
y			

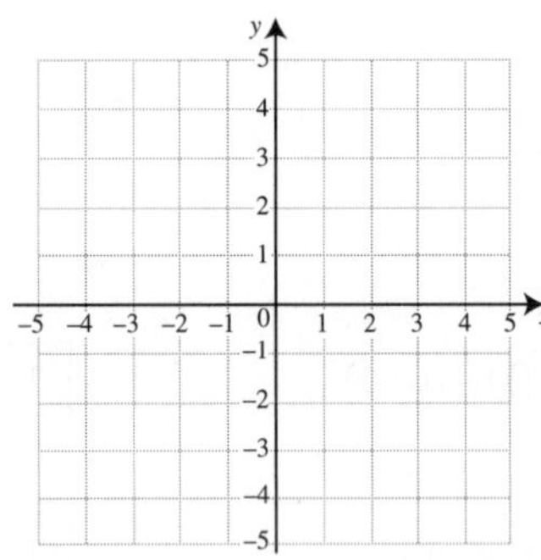

15 On the same number plane draw the graphs of $x = 3$ and $y = -4$, and find the point of intersection of the two lines. 3

16 The points $A(0, 1)$ and $B(4, 4)$ are plotted on the number plane. Find the:

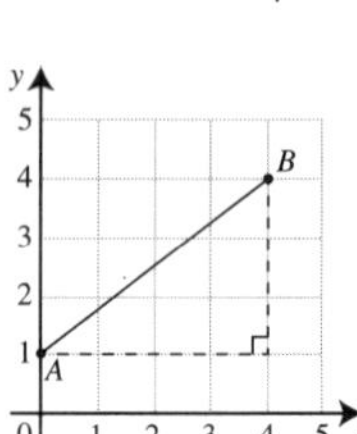

a length of AB **b** gradient of AB **c** midpoint of AB

____________ ____________ ____________

____________ ____________ ____________ 6

17 For the following equations write down the gradient (m) and the y-intercept (b):

a $y = 3x + 5$ **b** $y = 2 - 5x$ **c** $y = x$

m ________ m ________ m ________ 3

b ________ b ________ b ________ 3

18 What is the equation of the line with:

a gradient 2 and y-intercept 1? **b** gradient 3 and y-intercept -2? **c** gradient -1 and y-intercept 0?

____________ ____________ ____________ 3

Total marks

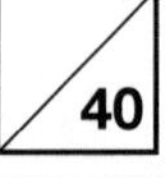

Linear Relationships

LEVEL 2 TEST — PART A

Time allowed: 40 minutes

Marks

1 Which of these is the equation of the line with gradient –2 and y-intercept 3?

(A) $y = 3 + 2x$ (B) $y = 3x - 2$ (C) $y = 2x - 3$ (D) $y = 3 - 2x$ 1

2 What is the midpoint of PQ if $P(0, 8)$ and $Q(6, 4)$?

(A) (6, 12) (B) (3, 6) (C) (3, 2) (D) (4, 5) 1

3 Which of these points does the line $2x - 3y = 6$ pass through?

(A) (0, 3) (B) (1, 1) (C) (6, 1) (D) (–3, –4) 1

4 Which of these lines is parallel to $y = 3x + 2$ and passes through the origin?

(A) $y = 3x - 2$ (B) $y = 3x$ (C) $y = 2$ (D) $y = 2x$ 1

5 What is the equation of the line parallel to the x-axis and passing through (2, –5)?

(A) $x = 2$ (B) $y = 2$ (C) $x = -5$ (D) $y = -5$ 1

6 Which of these lines is parallel to the line with equation $y = \dfrac{2x+1}{3}$?

(A) $y = 2x + 3$ (B) $2x - y = 1$ (C) $2x - 3y = 4$ (D) $3x - 2y = 1$ 1

The lines m and n are drawn on the number plane which is used to answer questions 7 to 13.

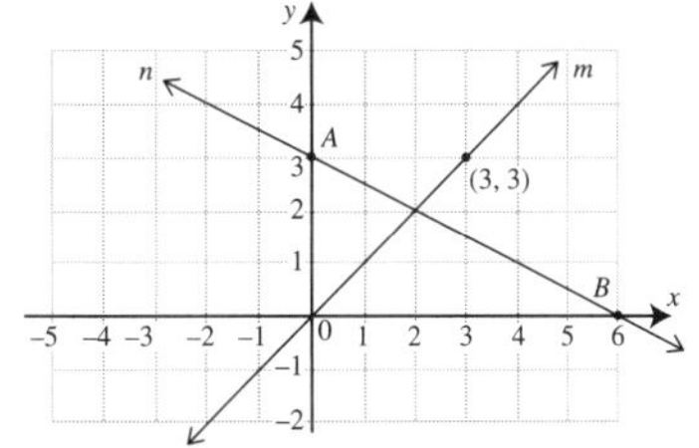

7 What is the gradient of the line n?

(A) –2 (B) $-\frac{1}{2}$ (C) $\frac{1}{2}$ (D) 2 1

8 Which of these is the gradient of the line m?

(A) –1 (B) 0 (C) 1 (D) 3 1

9 What is the distance from the origin to the point (3, 3)?

(A) 1 unit (B) 3 units (C) 6 units (D) $\sqrt{18}$ units 1

10 Which of these is the equation of line m?

(A) $y = x + 3$ (B) $y = 3x + 3$ (C) $y = x$ (D) $y = 3x$ 1

11 Which of these points would lie on the line m?

(A) (–2, –2) (B) (–2, –1) (C) (3, 4) (D) (4, 3) 1

12 What is the midpoint of the interval AB?

(A) $(3, 1\frac{1}{2})$ (B) $(1\frac{1}{2}, 3)$ (C) $(1\frac{1}{2}, 1\frac{1}{2})$ (D) (3, 3) 1

13 Which of these is the length of the interval joining the points A and B?

(A) $\sqrt{45}$ units (B) $\sqrt{65}$ units (C) 9 units (D) 11 units 1

LEVEL 2 TEST

PART B

Write answers and working in the space provided.

Marks

14 The points $P(-5, 1)$ and $Q(2, -3)$ are plotted on the number plane. Find the:

a length of PQ 2

____

____

b gradient of PQ 2

____

____

c midpoint of PQ 1

____

____

15 Find the gradient (m), the y-intercept (b) and the equation of the following lines:

a

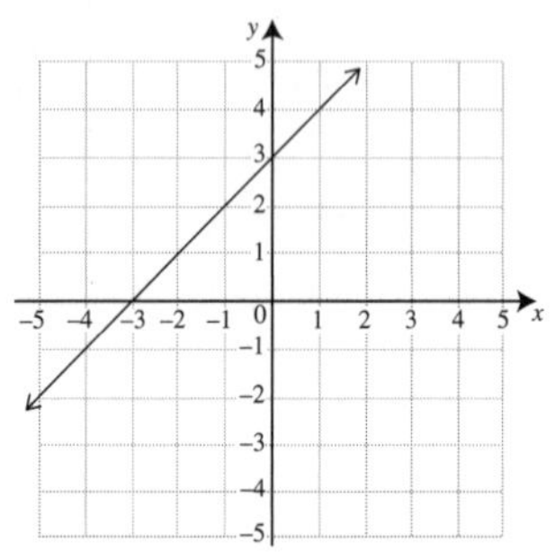

m: ______________________

b: ______________________

eqn: ______________________

b

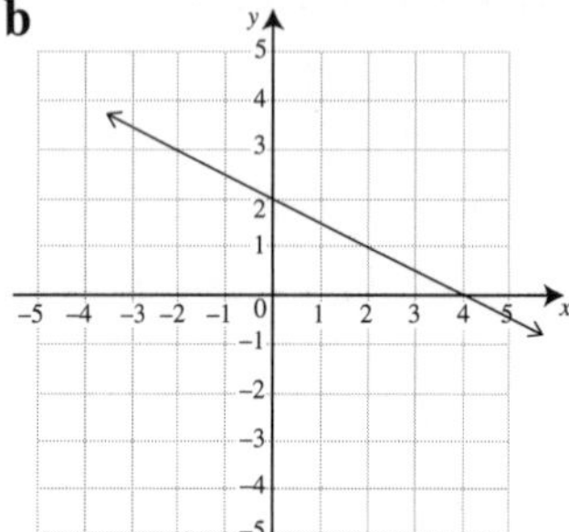

m: ______________________

b: ______________________

eqn: ______________________

c

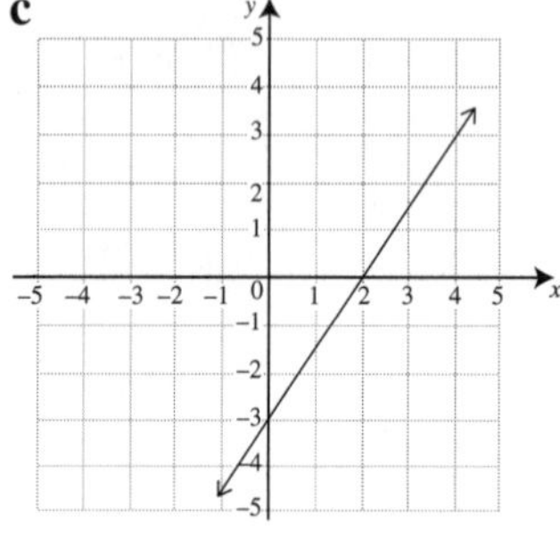

m: ______________________

b: ______________________

eqn: ______________________

3 3 3

16 Graph the pair of equations on a number plane and find the point of intersection. 9

a $y = 2x - 1$ and $y = 3 - 2x$ **b** $y = x + 1$ and $x + y = 3$ **c** $3x + y + 4 = 0$ and $y = x$

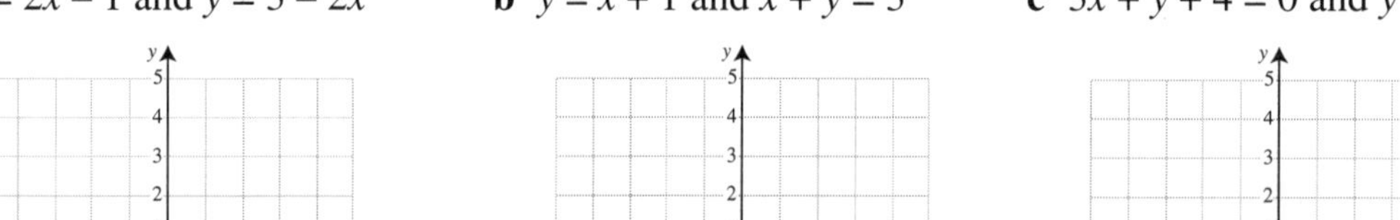

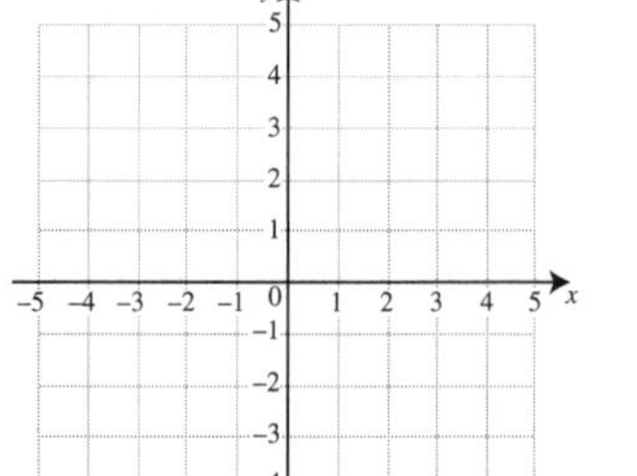

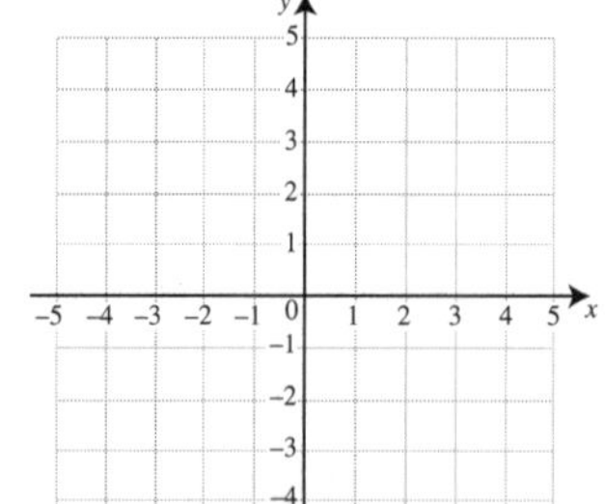

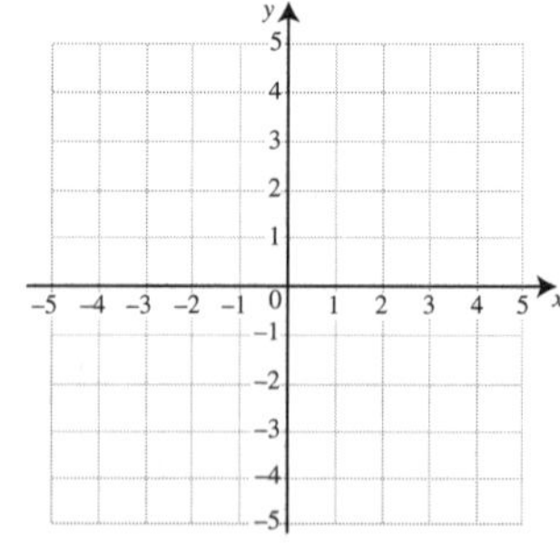

17 Without the use of a number plane, what is the point of intersection of the two lines?

a $x = -1$ and $y = 5x - 4$

____

____

b $x - 3y + 7 = 0$ and $y = 2$ 4

____

____

Total marks

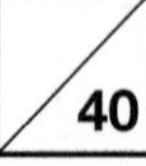

Linear Relationships

LEVEL 3 TEST — PART A

Time allowed: 40 minutes

Marks

1 The line $2x - 3y = 5$ passes through the point $(s, -1)$. What is the value of s?

Ⓐ $s = -1$ Ⓑ $s = 1$ Ⓒ $s = 2$ Ⓓ $s = 4$ 1

2 The points $O(0, 0)$, $P(2, 5)$, $Q(6, 5)$ and R are vertices of parallelogram $OPQR$. What are the coordinates of R?

Ⓐ $(-2, 0)$ Ⓑ $(-2, 0)$ Ⓒ $(4, 0)$ Ⓓ $(10, 0)$ 1

3 $M(5, 2)$ is the midpoint of AB, where $A(-1, 8)$. What are the coordinates of B?

Ⓐ $(2, 5)$ Ⓑ $(9, -4)$ Ⓒ $(11, -4)$ Ⓓ $(-7, -6)$ 1

4 Which of these is the gradient of the line with equation $2x - 4y + 3 = 0$?

Ⓐ -2 Ⓑ $-\frac{1}{2}$ Ⓒ $\frac{1}{2}$ Ⓓ 2 1

5 If $X(3, -1)$ and $Y(-1, 5)$, what is the exact length of XY?

Ⓐ $\sqrt{6}$ units Ⓑ 4 units Ⓒ $\sqrt{20}$ units Ⓓ $\sqrt{52}$ units 1

6 The gradient of the interval CD is 2. If $C(-4, 0)$, which of these are possible coordinates of the point D?

Ⓐ $(-2, 0)$ Ⓑ $(0, 2)$ Ⓒ $(0, -2)$ Ⓓ $(0, 8)$ 1

7 If the lines $ax - 2y - 3 = 0$ and $y = 4 - 2x$ are parallel, what is the value of a?

Ⓐ $a = -4$ Ⓑ $a = -1$ Ⓒ $a = 1$ Ⓓ $a = 4$ 1

8 The line $x - y + 2 = 0$ cuts the axes at A and B. What is the length of AB?

Ⓐ 1 unit Ⓑ 2 units Ⓒ $\sqrt{8}$ units Ⓓ 4 units 1

9 Which of these is the point of intersection of $y = x - 1$ and $y = 3x + 5$?

Ⓐ $(-1, 2)$ Ⓑ $(2, 1)$ Ⓒ $(3, 2)$ Ⓓ $(-3, -4)$ 1

10 Which of these is parallel to the line $2x - 4y = 1$?

Ⓐ $y = \frac{x-3}{2}$ Ⓑ $y = 2x + 5$ Ⓒ $4x - 2y = 1$ Ⓓ $y = \frac{1-x}{2}$ 1

11 A line parallel to $y = 3 - 2x$ passes through $(1, 3)$. What is the equation of the line?

Ⓐ $x + 2y - 7 = 0$ Ⓑ $3x + y - 6 = 0$ Ⓒ $4x + 2y - 7 = 0$ Ⓓ $2x + y - 5 = 0$ 1

12 The line $3x + 4y = 12$ cuts the axes at S and T. If O is the origin, which of these is the perimeter of the triangle SOT?

Ⓐ 12 units Ⓑ 15 units Ⓒ 16 units Ⓓ 19 units 1

13 The line $y = ax + 4$ passes through the point $(-1, 2)$. What are the coordinates of the midpoint of the interval joining the x-intercept and y-intercept?

Ⓐ $(-1, 2)$ Ⓑ $(1, 2)$ Ⓒ $(2, 1)$ Ⓓ $(2, -1)$ 1

LEVEL 3 TEST PART B

Write answers and working in the space provided.

Marks

14 Graph the following lines: [9]

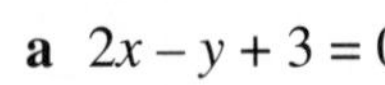

a $2x - y + 3 = 0$ **b** $y = \dfrac{3x - 1}{2}$ **c** $3x + 2y + 1 = 0$

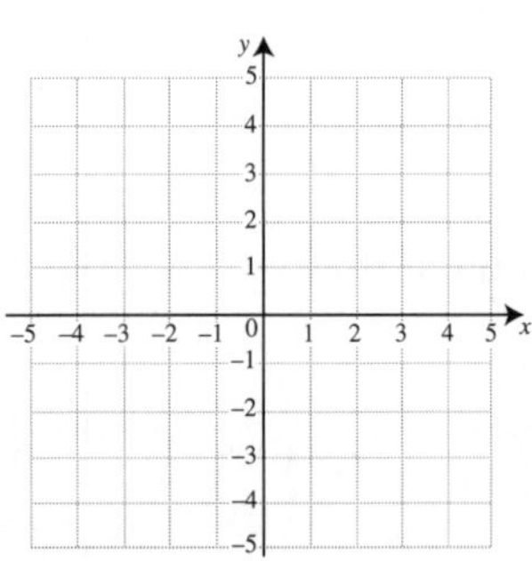

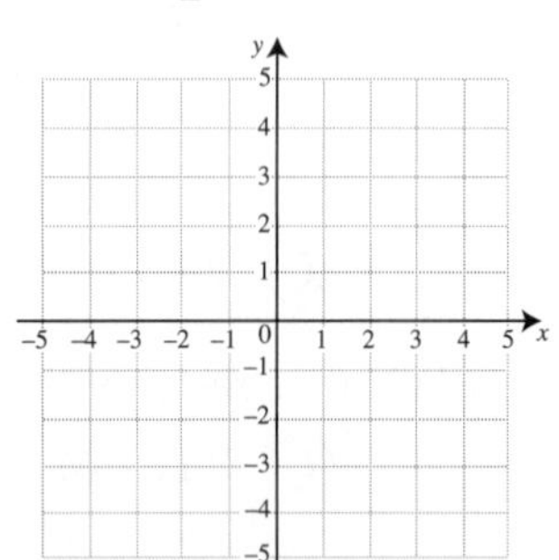

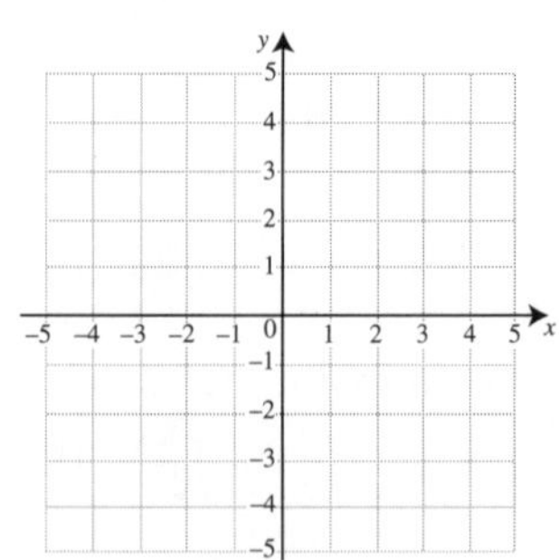

15 If $A(-2, 5)$ and $B(3, -1)$, find:

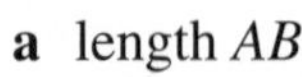

a length AB

b midpoint AB

c gradient AB

[6]

16 Find the point of intersection of the following pairs of lines:

a $y = 3x - 2$ and $y = 5x - 4$ **b** $x - 3y + 10 = 0$ and $y = 2x$

[6]

17 Find the equation of the line which:

a has a y-intercept of 2 and passes through $(3, -4)$

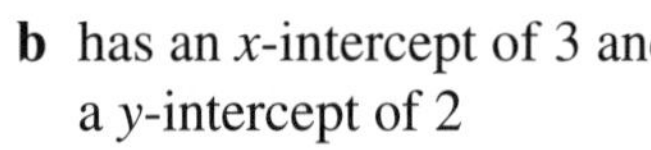

b has an x-intercept of 3 and a y-intercept of 2

c passes through $(-1, 2)$ and is parallel to $2x - y + 1 = 0$

[6]

Total marks

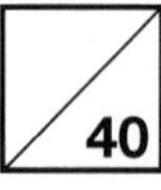

/40

Quadratics

PRETEST LEVEL 1

Time allowed: 10 minutes **Total marks: 10**

Marks

1 Simplify $x^2 + 3x + 2x + 6$.

(A) $6x^2 + 6$ (B) $12x^2$ (C) 12 (D) $x^2 + 5x + 6$ 1

2 Solve $a + 5 = 0$.

(A) $a = -5$ (B) $a = -1$ (C) $a = 1$ (D) $a = 5$ 1

3 If $2p = 7$, what is the value of p?

(A) $-3\frac{1}{2}$ (B) $3\frac{1}{2}$ (C) 5 (D) 9 1

4 Which of these is the expansion of $2(m + 6)$?

(A) $2m + 6$ (B) $2m + 8$ (C) $2m + 12$ (D) $12m$ 1

5 Solve the equation $x^2 = 16$.

(A) $x = 4$ (B) $x = 8$ (C) $x = \pm 4$ (D) $x = \pm 8$ 1

6 Expand $a(a - 7)$.

(A) $2a - 7$ (B) $-6a$ (C) $a^2 - 7$ (D) $a^2 - 7a$ 1

7 The table is completed for the rule $y = x^2 + 1$. What are the values of p and q?

x	−2	−1	0	1	2
y	p	2	q	2	5

(A) $p = 5$ and $q = 1$ (B) $p = -5$ and $q = 1$ (C) $p = 5$ and $q = 0$ (D) $p = -5$ and $q = 0$ 1

8 What is the solution of the equation $x - 2 = 0$?

(A) $x = 2$ (B) $x = -2$ (C) $x = 0$ (D) $x = \frac{1}{2}$ 1

9 Simplify $d^2 - 2d + 2d - 4$.

(A) $d^2 + 4$ (B) $d^2 - 4$ (C) $d^2 - 4d$ (D) $-d^2$ 1

10 Expand $3(a + 2) + 1$.

(A) $3a + 6$ (B) $6a + 1$ (C) $3a + 7$ (D) $6a + 6$ 1

Total marks /10

Quadratics

PRETEST — LEVELS 2 AND 3

Time allowed: 10 minutes **Total marks: 10**

Marks

1 Which of these is the solution of $2x - 1 = 0$?

(A) $x = -\frac{1}{2}$ (B) $x = \frac{1}{2}$ (C) $x = 1$ (D) $x = 2$ — 1

2 Expand $m(m + 1) + 3(m + 1)$.

(A) $8m^2$ (B) $m^2 + 3m + 1$ (C) $m^2 + 4m + 3$ (D) $m^2 + 3m + 4$ — 1

3 Which of these is equal to $\frac{4a^2}{a}$?

(A) $2a$ (B) $2a$ (C) $4a$ (D) $\frac{4}{a}$ — 1

4 Expand $-(2p - 5)$.

(A) $-2p + 5$ (B) $-2p - 5$ (C) $3p$ (D) $-3p$ — 1

5 Factorise $6c^2 + 4c$.

(A) $6c(c + 4)$ (B) $3c(2c + 1)$ (C) $2c(c + 2)$ (D) $2c(3c + 2)$ — 1

6 Which of these is the same as $\frac{2p}{q+3} \times \frac{q+3}{6p^2}$?

(A) $\frac{1}{3p}$ (B) $\frac{1}{3p^2}$ (C) $\frac{2}{3p}$ (D) $\frac{1}{4p}$ — 1

7 Which of these is the solution of $4x^2 = 64$?

(A) $x = 8$ (B) $x = 4$ (C) $x = \pm 8$ (D) $x = \pm 4$ — 1

8 Lee thinks of two numbers that have a sum of –6 and a product of –16. What are the two numbers?

(A) –4 and –2 (B) –8 and 2 (C) –4 and –4 (D) 8 and –2 — 1

9 If $d = 1$, which of these is the value of $(a + 3)(b - 2)(c + 5)(d - 1)$?

(A) abc (B) $abc - 30$ (C) $abc + 30$ (D) 0 — 1

10 Which of these is **not** a factor of $3x^2 - 6x$?

(A) x (B) $3x^2$ (C) $x - 2$ (D) $x(x - 2)$ — 1

Total marks

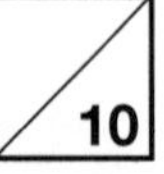

Quadratics

LEVEL 1 TEST PART A

Time allowed: 40 minutes

Marks

1 Expand $w(w-4)$.

(A) $2w-4$ (B) w^2-4 (C) w^2-4w (D) $-3w$ 1

2 What is the expansion of $2t(3t-5)$?

(A) $5t^2-10t$ (B) $6t^2-10t$ (C) $5t^2-3t$ (D) $6t^2-3t$ 1

3 What is the missing term in the result $(p+4)(p+2) = \ldots + 2p + 4p + 8$?

(A) p^2 (B) $2p$ (C) $6p$ (D) $8p$ 1

4 What is the missing term in the result $(m+5)(m-3) = m^2 - \ldots + 5m - 15$?

(A) $3m$ (B) $5m$ (C) $2m$ (D) m 1

5 What is the missing term in the result $(y-2)(y+3) = y^2 + 3y - \ldots - 6$?

(A) y (B) $2y$ (C) $5y$ (D) $6y$ 1

6 What is the missing term in the result $(a-4)(a-5) = a^2 - 5a - 4a + \ldots$?

(A) 1 (B) 2 (C) 9 (D) 20 1

7 Which of these pairs of numbers has a sum of 8 and a product of 12?

(A) 7 and 1 (B) 3 and 4 (C) 4 and 2 (D) 6 and 2 1

8 Solve the equation $(x-1)(x+3)=0$.

(A) $x=-1, 3$ (B) $x=1, -3$ (C) $x=1, 3$ (D) $x=-1, -3$ 1

9 If $2p^2 = 18$, what is the value of p?

(A) $p=9$ (B) $p=3$ (C) $p=\pm 9$ (D) $p=\pm 3$ 1

10

$(2x+5)$ cm

$(x+3)$ cm

Which of these is an expression for the area of the rectangle?

(A) $(3x+8)$ cm^2 (B) $4(2x+5)$ cm^2 (C) $(2x+5)(x+3)$ cm^2 (D) $4(x+3)$ cm^2 

11 Solve the equation $m(m+5)=0$.

(A) $m=5$ (B) $m=-5$ (C) $m=0, 5$ (D) $m=0, -5$ 1

12 Factorise p^2-4p.

(A) $p(p-4)$ (B) $p(p-2)$ (C) $(p-2)(p-2)$ (D) $2p(p-2)$ 1

13 Which of these is the solution of $g^2+8=9$?

(A) $g=1$ (B) $g=\pm 1$ (C) $g=\pm 3$ (D) $g=-5$ 1

Quadratics

LEVEL 1 TEST — PART B

Write answers and working in the space provided.

Marks

14 Expand:

a $(x + 3)(x + 5)$ b $(y + 6)(y - 1)$ c $(a - 1)(a - 2)$

3

15 Expand:

a $(2y + 1)(y + 3)$ b $(5p - 1)(p + 2)$ c $(3m - 2)(2m - 1)$

6

16 Factorise:

a $a^2 + 3a + 2$ b $w^2 + 6w + 5$ c $b^2 - 3b + 2$

3

d $x^2 + 4x - 5$ e $y^2 - y - 6$ f $z^2 - 8z + 12$

3

17 Solve:

a $(x + 1)(x - 2) = 0$ b $(y + 4)(y - 5) = 0$ c $m(m - 2) = 0$

3

d $(c + 3)(c - 3) = 0$ e $(a - 5)(a + 5) = 0$ f $2y(y - 3) = 0$

3

18 Solve:

a $b^2 + 4b + 3 = 0$ b $n^2 - 7n + 12 = 0$ c $h^2 + 3h - 10 = 0$

6

Total marks

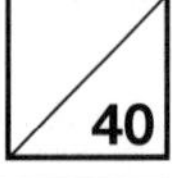

/40

Quadratics

LEVEL 2 TEST — PART A

Time allowed: 40 minutes

Marks

1 Which of these is the expansion of $(x-3)(x+5)$?

(A) $x^2+2x-15$ (B) $x^2-15x+2$ (C) $x^2-2x+15$ (D) $x^2-2x-15$ — 1

2 Expand $(a-5)^2$.

(A) a^2-10 (B) a^2-25 (C) $a^2-5a+25$ (D) $a^2-10a+25$ — 1

3 Which of these is the solution of the equation $m^2-2m-8=0$?

(A) $m=-2,-4$ (B) $m=-2, 4$ (C) $m=2,-4$ (D) $m=2, 4$ — 1

4 A square has a side length of $(x+3)$ cm.
Which of these is the area of the square?

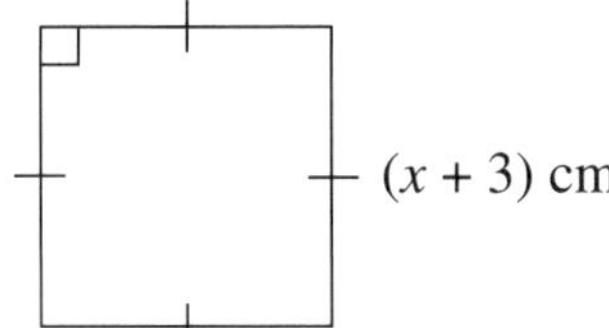

(A) $(4x+12)$ cm² (B) (x^2+9) cm² (C) (x^2+3x+9) cm² (D) (x^2+6x+9) cm² — 1

5 Solve the quadratic equation $4p^2=400$.

(A) $p=10$ (B) $p=20$ (C) $p=\pm10$ (D) $p=\pm20$ — 1

6 If $(x+a)(x+b)=x^2-10x+16$, what are the values of a and b?

(A) 4 and 4 (B) –2 and 8 (C) 2 and –8 (D) –2 and –8 — 1

7 Expand and simplify $(2w-1)(w+5)$.

(A) $2w^2+10w-5$ (B) $2w^2+4w+5$ (C) $2w^2+9w-5$ (D) $2w^2+8w-5$ — 1

8 Factorise $a(m-n)+b(m-n)$.

(A) $(m+n)(a-b)$ (B) $(m-n)(a+b)$ (C) $(a-m)(b+n)$ (D) $(a+m)(b-n)$ — 1

9 Solve the equation $4-16y^2=0$.

(A) $y=\pm\frac{1}{4}$ (B) $y=\pm\frac{1}{2}$ (C) $y=\pm2$ (D) $y=\pm4$ — 1

10 Simplify $\dfrac{2x-4}{(x-2)(x+2)}$.

(A) $\dfrac{2}{x+2}$ (B) $\dfrac{1}{x+1}$ (C) $\dfrac{2x-1}{(x-1)(x+1)}$ (D) $\dfrac{2}{x+1}$ — 1

11 Which of these is the solution of $t^2+t-20=0$?

(A) $t=-2, 10$ (B) $t=2,-10$ (C) $t=4,-5$ (D) $t=-4, 5$ — 1

12 Simplify $\dfrac{x-1}{x^2-x}$.

(A) $\dfrac{-1}{x^2}$ (B) $\dfrac{-1}{2}$ (C) $\dfrac{1}{x}$ (D) x — 1

13 Factorise the expression $3p^2-3p-6$.

(A) $(p+2)(p-3)$ (B) $(p+1)(p-2)$ (C) $(p+1)(p-6)$ (D) $3(p+1)(p-2)$ — 1

Quadratics

LEVEL 2 TEST — PART B

Write answers and working in the space provided.

Marks

14 Expand and simplify:

a $(2a - 1)(3a + 4)$

b $(4y + 3)(2y - 1)$

c $(3p - 2)(4p - 5)$

3

15 Factorise:

a $p^2 + 3p - 3pq - 9q$

b $m^2 - 5m - 3mn + 15n$

c $2y^2 + 12yz - 3y - 18z$

6

16 Factorise:

a $2b^2 - 2b - 24$

b $3x^2 - 24x + 45$

c $5q^2 - 30q + 40$

6

17 Solve the following equations:

a $n^2 - 5n - 6 = 0$

b $g^2 - 11g + 18 = 0$

c $p^2 - 3p = 10$

6

18 Simplify the following:

a $\dfrac{x^2 - 9}{2x - 6}$

b $\dfrac{4y - 8}{y^2 - 4y + 4}$

c $\dfrac{2m^2 - 2}{2m - 2}$

6

Total marks 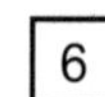 40

Quadratics

LEVEL 3 TEST — PART A

Time allowed: 40 minutes

Marks

1 Factorise $x(x-y)+y(y-x)$.

Ⓐ $(x-y)^2$ Ⓑ $(x-2y)^2$ Ⓒ $(x-y)(x+y)$ Ⓓ $(x+y)^2$ — 1

2 Expand and simplify $(a+b)^2-(a-b)^2$.

Ⓐ $4ab$ Ⓑ $2ab$ Ⓒ $2a^2-2b^2$ Ⓓ $2a^2$ — 1

3 Which of these is equal to the expansion of $(p+\frac{1}{p})^2$?

Ⓐ 1 Ⓑ $p^2+\frac{1}{p^2}$ Ⓒ $p^2+2p+\frac{1}{p^2}$ Ⓓ $p^2+2+\frac{1}{p^2}$ — 1

4 Factorise $c(c-5)+10-2c$.

Ⓐ $(c-5)(c-2)$ Ⓑ $(c-5)(c+2)$ Ⓒ $(c+5)(c-2)$ Ⓓ $(c+5)(c+2)$ — 1

5 Simplify $\frac{b^2-9}{3-b}$.

Ⓐ $b+3$ Ⓑ $-b-3$ Ⓒ $b-3$ Ⓓ $-b+3$ — 1

6 Factorise $3x^2+13x-10$.

Ⓐ $(3x+2)(x-5)$ Ⓑ $(3x-5)(x+2)$ Ⓒ $(3x-2)(x+5)$ Ⓓ $(3x+5)(x-2)$ — 1

7 Simplify $\frac{x^2}{x^2-x}$.

Ⓐ $\frac{1}{1-x}$ Ⓑ $\frac{1}{x-1}$ Ⓒ $\frac{x}{x-1}$ Ⓓ $\frac{1}{x}$ — 1

8 Solve $2x^2-6x-8=0$.

Ⓐ $x=2, 4$ Ⓑ $x=4, -1$ Ⓒ $x=4, 1$ Ⓓ $x=2, 4, -1$ — 1

9 Solve $a^3+2a^2+a=0$.

Ⓐ $a=0, 1$ Ⓑ $a=0, -1$ Ⓒ $a=2$ Ⓓ $a=3$ — 1

10 Which of these are the solutions of the equation $2x^2=x$?

Ⓐ $x=0, 2$ Ⓑ $x=0, -2$ Ⓒ $x=0, -\frac{1}{2}$ Ⓓ $x=0, \frac{1}{2}$ — 1

11 The product of two consecutive integers is 12. Which of these could **not** be one of the numbers?

Ⓐ -4 Ⓑ -3 Ⓒ 3 Ⓓ 6 — 1

12 Factorise $x^2-(y+2)^2$.

Ⓐ $(x-y+2)(x+y+2)$ Ⓑ $(x-y+2)(x-y-2)$

Ⓒ $(x-y-2)(x+y+2)$ Ⓓ $(x-y-2)(x+y-2)$ — 1

13 Simplify $\frac{2}{x}-\frac{1}{x^2}$.

Ⓐ $\frac{1}{x^3}$ Ⓑ $\frac{1}{x^2}$ Ⓒ $\frac{2-x}{x^2}$ Ⓓ $\frac{2x-1}{x^2}$ — 1

LEVEL 3 TEST

PART B

Write answers and working in the space provided.

Marks

14 Expand and simplify:

a $(2w + 3)^2$

b $(3y - 1)^2$

c $(5 - 2t)(5 + 2t)$

3

15 Factorise:

a $3d^2 - 2d - 8$

b $2x^2 + 5x - 3$

c $6p^2 - 11p + 4$

6

16 Solve:

a $2m^2 - 5m - 12 = 0$

b $4v^2 - 21v + 5 = 0$

c $10q^2 + 9q - 9 = 0$

6

17 Simplify:

a $\dfrac{a^2 - 3a - 4}{a^2 - 1} \times \dfrac{a + 1}{a - 4}$

b $\dfrac{x^2 - 4}{x^2 + 5x + 6} \div \dfrac{x - 2}{x + 3}$

6

18 Solve simultaneously the equations:

a $y = x^2 + 3x - 4$ and $y = 2x + 2$

b $y = 2x^2 + 3x - 5$ and $y = x^2 + 5x + 3$

6

Total marks  40

Rates and Proportion

PRETEST LEVEL 1

Time allowed: 10 minutes **Total marks: 10**

Marks

1 Chloe's heart beats 80 times per minute. How many times will it beat in one hour?

Ⓐ 140 Ⓑ 480 Ⓒ 800 Ⓓ 4800 1

2 Jarrod's car uses petrol at the rate of 7.8 L/100 km on a journey of 500 km. What was the total quantity of fuel used?

Ⓐ 39 L Ⓑ 48 L Ⓒ 50 L Ⓓ 3900 L 1

3 In a bag there are three red balls for every four green balls. If the bag contains 12 green balls, how many red balls are in the bag?

Ⓐ 6 Ⓑ 8 Ⓒ 9 Ⓓ 36 1

4 When Brad worked on Saturday night he was paid $163.20 for 6 hours work. Which of these was Brad's hourly rate?

Ⓐ $27.20 Ⓑ $29.40 Ⓒ $169.20 Ⓓ $979.20 1

5 Clare cycled for 30 minutes and covered a distance of 12 km. What was Clare's average speed for the ride?

Ⓐ 6 km/h Ⓑ 12 km/h Ⓒ 18 km/h Ⓓ 24 km/h 1

6 Carolyn places two stamps on every padded envelope she posts in the mail. If each stamp costs $1.10, what is the total cost of stamps when she posts 16 envelopes?

Ⓐ $2.20 Ⓑ $17.60 Ⓒ $32 Ⓓ $35.20 1

7 Justin lives 8 km from where he works and each day he drives the same route. Last week he worked for 5 days. What was the total distance travelled?

Ⓐ 40 km Ⓑ 80 km Ⓒ 112 km Ⓓ 120 km 1

8 The price of petrol is advertised at 159.9 cents/L. If William purchased 45.8 L of petrol and received a discount of 4 cents/L, how much did the petrol cost, to the nearest cent?

Ⓐ $66.84 Ⓑ $71.40 Ⓒ $73.05 Ⓓ $73.23 1

9 If $A = 2m$, which of these is the value of m when $A = 8$?

Ⓐ $m = 16$ Ⓑ $m = 10$ Ⓒ $m = 6$ Ⓓ $m = 4$ 1

10 It takes three workers 2 hours to dig a 12-metre long ditch. How long would it have taken six workers to dig the same ditch?

Ⓐ 30 minutes Ⓑ 1 hour Ⓒ 4 hours Ⓓ 6 hours 1

Total marks /10

Rates and Proportion

PRETEST — LEVELS 2 AND 3

Time allowed: 10 minutes — **Total marks: 10**

Marks

1 If $P = mv^2$, what is the value of P when $m = 0.5$ and $v = 4$?

(A) 4 (B) 8 (C) 12 (D) 24 — 1

2 Which of these equations represents a straight line on a number plane?

(A) $y = 0.2x$ (B) $y = x^2$ (C) $y = \frac{2}{x}$ (D) $y = 2^x$ — 1

3 When $y = 0.6x$, which of these is the value of x when $y = 12$?

(A) $x = 0.05$ (B) $x = 0.2$ (C) $x = 7.2$ (D) $x = 20$ — 1

4 A cube with a volume of 1 m^3 is produced. How many cubic centimetres are in the cube?

(A) 100 (B) 1000 (C) 10 000 (D) 1 000 000 — 1

5 If $m = \frac{2}{\sqrt{n}}$, what is the value of n when $m = 5$?

(A) $n = 0.04$ (B) $n = 0.16$ (C) $n = 0.4$ (D) $n = 100$ — 1

6 A car is travelling at 60 km/h. How far will it travel in one minute?

(A) 1 m (B) 10 m (C) 100 m (D) 1000 m — 1

The graph shows the distance of a car from home and is used to answer questions 7 to 10.

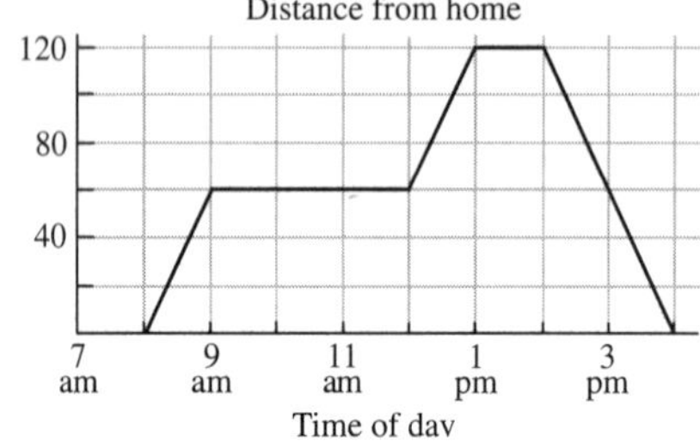

7 Which of these is the total distance travelled by the car during the day?

(A) 500 km (B) 70 km (C) 120 km (D) 240 km — 1

8 What was the longest time that the car was stationary?

(A) 2 hours (B) 2.5 hours (C) 3 hours (D) 8 hours — 1

9 How far did the car travel between 10 am and 1 pm?

(A) 40 km (B) 60 km (C) 80 km (D) 120 km — 1

10 Which of these is the average speed of the car between 2 pm and 4 pm?

(A) 60 km/h (B) 80 km/h (C) 90 km/h (D) 120 km/h — 1

Total marks

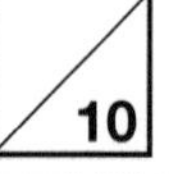

Rates and Proportion

LEVEL 1 TEST — PART A

Time allowed: 40 minutes

Marks

The graph shows the distance from home of a motorist and is used to answer questions 1 to 5.

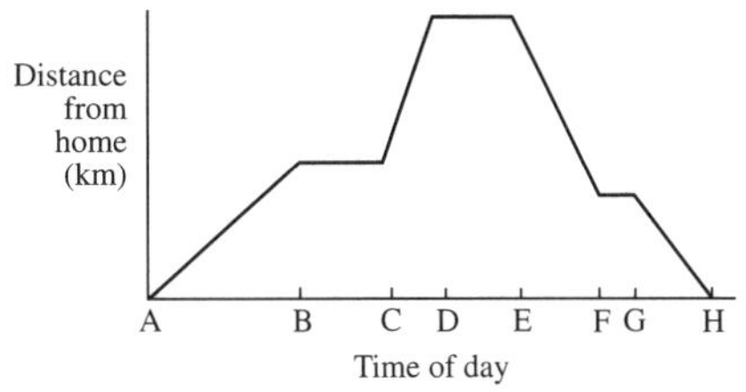

Distance from home (km)

A B C D E F G H

Time of day

1 At what time did the motorist arrive back home?

(A) *A* (B) *C* (C) *D* (D) *H* — 1

2 Between which two times was the motorist away from home?

(A) *A* and *B* (B) *B* and *C* (C) *A* and *D* (D) *A* and *H* — 1

3 How many times did the motorist stop?

(A) 1 (B) 2 (C) 3 (D) 4 — 1

4 Between which two times was the motorist travelling the fastest?

(A) *B* and *C* (B) *C* and *D* (C) *F* and *H* (D) *G* and *H* — 1

5 Between which two times did the motorist stop for the shortest time?

(A) *B* and *C* (B) *C* and *D* (C) *F* and *G* (D) *A* and *H* — 1

6 Which of these is the same rate as 60 L in 4 hours?

(A) 15 L/h (B) 56 L/h (C) 64 L/h (D) 240 L/h — 1

7 If $D = ST$, which of these is the value of D when $S = 80$ and $T = 4$?

(A) 20 (B) 84 (C) 160 (D) 320 — 1

8 A recipe lists three eggs to make eight cakes. How many eggs are required to make 24 cakes?

(A) 6 (B) 9 (C) 12 (D) 19 — 1

9 Which of these is the same speed as 60 km/h?

(A) 1 m/s (B) 10 m/s (C) 10 km/min (D) 1 km/min — 1

10 The cost C, in \$, of a length of pipe n metres is calculated using the formula $C = 7.5n$. What is the cost of 8.5 metres of pipe?

(A) \$18 (B) \$63.75 (C) \$65.75 (D) \$68.75 — 1

11 If $y = kx$, what is the value of k when $y = 8$ and $x = 4$?

(A) $k = 2$ (B) $k = 4$ (C) $k = 16$ (D) $k = 32$ — 1

12 Which of these is the value of p if $\frac{12}{p} = \frac{2}{3}$?

(A) $p = 4$ (B) $p = 6$ (C) $p = 16$ (D) $p = 18$ — 1

13 If $C = kA$, what is the value of A when $C = 480$ and $k = 2.4$?

(A) $A = 200$ (B) $A = 240$ (C) $A = 960$ (D) $A = 1152$ — 1

Rates and Proportion

LEVEL 1 TEST — PART B

Write answers and working in the space provided.

Marks

14 This table links the values of y and x.

x	2	4	6	B	10
y	12	24	A	54	60

a Complete the rule: $y =$ .... **b** Find the value of A. **c** Find the value of B.

3

15 If $M = 8T$ find:

a M when $T = 4.2$ **b** M when $T = 120$ **c** T when $M = 48$

6

16 Kendall shows the relationship between A and t by writing the equation $A = kt$. Find:

a A, when $k = 0.2$ and $t = 3$ **b** k, when $A = 16$ and $t = 8$ **c** t, when $A = 48$ and $k = 10$

6

17 Use the pronumeral k to write an equation relating the two pronumerals, if:

a P is directly proportional to n **b** T is directly proportional to s **c** d is directly proportional to A

3

18 Using the pronumeral k, write an equation that links the two pronumerals if the:

a cost (C) of a toll is directly proportional to the length of a road (L)

b length of a ditch (L) is directly proportional to the length of time digging (t)

c cost of painting a cube (C) is directly proportional to the square of the length of each side (s)

3

19 To paint an area of 80 m^2, Anna uses 10 L of paint.

a Write a formula linking area (A) and amount of paint (p) in litres.

b What area will she paint if she uses 20 L?

c How much paint will Anna use to paint 120 m^2?

6

Total marks

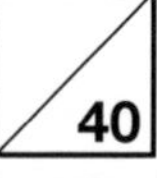

Rates and Proportion

LEVEL 2 TEST PART A

Time allowed: 40 minutes

Marks

1 If P is directly proportional to t, which of the following equations is correct?

(A) $P = kt$ (B) $P = t - k$ (C) $P = k + t$ (D) $P = \frac{k}{t}$ 1

2 If y is directly proportional to the square of x, which of the following equations is correct?

(A) $y = \frac{2k}{x}$ (B) $y = \sqrt{k}\,x^2$ (C) $y = kx^2$ (D) $y = \sqrt{kx}$ 1

3 Which of these speeds is the closest to 60 km/h?

(A) 10 m/s (B) 12.1 m/s (C) 15.2 m/s (D) 16.7 m/s 1

4 If $M = 8$ when $n = 2$ and M is directly proportional to n, what is the constant of proportionality?

(A) 2 (B) 4 (C) 6 (D) 10 1

5 A motorist leaves home and drives at 80 km/h for one hour, 100 km/h for half an hour and 60 km/h for another half an hour. She then stops for one hour. Which of these graphs best represents her trip?

(A)
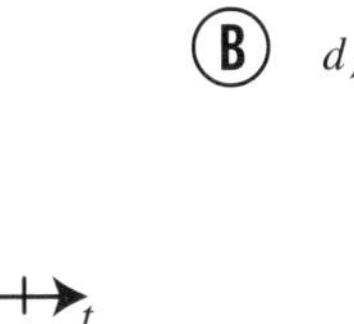

(B)
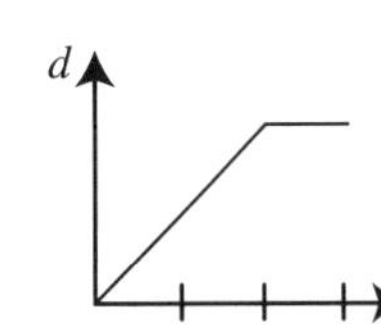

(C) (D) 1

6 Which of these is the same rate as $80c/m^2$?

(A) \$80/ha (B) \$800/ha (C) \$8000/ha (D) \$80 000/ha 1

The height h cm of a plant is directly proportional to the number of days d it has been growing. The plant is 20 cm tall after 8 days growth. This information is used to answer questions 7 to 10.

7 If $h = kd$, which of these is the value of proportional constant k?

(A) $k = 0.2$ (B) $k = 0.25$ (C) $k = 2.25$ (D) $k = 2.5$ 1

8 What is the height of the plant after 12 days?

(A) 30 cm (B) 32 cm (C) 48 cm (D) 240 cm 1

9 How many days will it take for the plant to be 50 cm tall?

(A) 16 days (B) 20 days (C) 24 days (D) 25 days 1

10 In how many **more** days will the plant double its height of 20 cm?

(A) 4 days (B) 8 days (C) 12 days (D) 16 days 1

11 Which of these is the same rate as $\$0.40/cm^3$?

(A) $\$4/m^3$ (B) $\$40/m^3$ (C) $\$4000/m^3$ (D) $\$400\,000/m^3$ 1

12 Which of the following represents an inverse proportion?

(A) $y = 2x$ (B) $y = \frac{x}{2}$ (C) $y = \frac{2}{x}$ (D) $y = x - 2$ 1

13 If six workers take 4 hours to build a fence, how long will it take eight workers?

(A) 3 hours (B) 3 hours 20 minutes (C) 3 hours 30 minutes (D) 3 hours 40 minutes 1

LEVEL 2 TEST

PART B

Write answers and working in the space provided.

Marks

14 The graph shows that y is directly proportional to x.

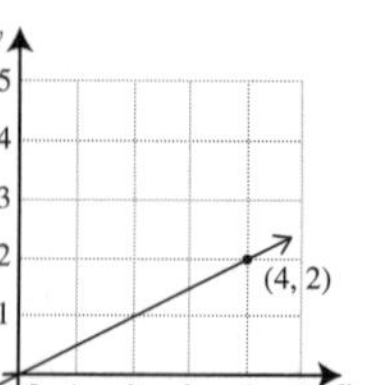

a Find the constant of proportionality.

b Write an equation that links y and x.

3

c Find the value of y if $x = 8$.

d Find the value of x if $y = 6$.

3

15 Olivia uses a fitness tracker to record her daily walking steps s and distance d in metres. On one day she recorded 9600 steps for a distance of 7.68 km.

a Determine an equation for this relationship in the form $d = ks$.

b How far will Olivia walk if she completes 12 400 steps?

c How many steps are required to record a distance of 6 km?

6

16 P is directly proportional to L^2 and when $L = 3$, $P = 108$.

a Write a formula linking P with L.

b What is the value of P when $L = 1.5$?

c If $P = 491.52$, find the value of positive L.

6

17 The mass m g of a ball is directly proportional to the cube of its radius r cm. A ball with mass 50 g has a radius of 2.5 cm.

a Write a formula linking m with r.

b If a ball has a mass of 137.2 g, what is the radius?

c What is the mass of a ball with diameter 12 cm?

6

18 The amount of varnish V needed to cover a wooden surface is directly proportional to the area A of the surface. If it takes 1.8 L to cover an area of 4 m^2, what quantity of varnish is required to cover a rectangle measuring 3 m by 2 m?

3

Total marks

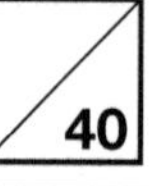

Rates and Proportion

LEVEL 3 TEST — PART A

Time allowed: 40 minutes

Marks

1 P is inversely proportional to the square root of n and $n = 9$ when $P = 4$.

Which of these is the equation which links P and n?

Ⓐ $n = \frac{12}{\sqrt{P}}$ Ⓑ $P = \frac{12}{\sqrt{n}}$ Ⓒ $P = \frac{9}{2\sqrt{n}}$ Ⓓ $n = \sqrt{\frac{324}{P}}$ — 1

The distance d a spring stretches when an object is attached to it is directly proportional to the mass m of the object. A mass of 56 g stretches a spring by 4.48 cm. This information is used to answer questions 2 to 4.

2 Which of these is an equation that links d and m?

Ⓐ $d = 0.08m$ Ⓑ $d = 12.5m$ Ⓒ $d = 51.52m$ Ⓓ $d = 60.48m$ — 1

3 How far is the spring stretched by a mass of 86 g?

Ⓐ 5.48 cm Ⓑ 6.28 cm Ⓒ 6.68 cm Ⓓ 6.88 cm — 1

4 Which of these is the mass of an object that stretches the string 10 cm?

Ⓐ 112 g Ⓑ 116 g Ⓒ 120 g Ⓓ 125 g — 1

The height h reached by an object thrown into the air varies in direct proportion to the square of the speed s at which it is thrown. An object thrown at a speed of 5 m/s reaches a height of 80 m. This information is used to answer questions 5 to 7.

5 Which of these is an equation that links h and s?

Ⓐ $h = 16s^2$ Ⓑ $h = 10.24s^2$ Ⓒ $h = 8s^2$ Ⓓ $h = 3.2s^2$ — 1

6 At what speed will an object be thrown to reach a height of 51.2 m?

Ⓐ 4 m/s Ⓑ 4.1 m/s Ⓒ 4.2 m/s Ⓓ 4.4 m/s — 1

7 Which of these is the maximum height the object reaches if thrown at 6 m/s?

Ⓐ 112.2 m Ⓑ 115.2 m Ⓒ 116.8 m Ⓓ 117.4 m — 1

The intensity of light L is inversely proportional to the square of the distance d from the source. At a distance of 25 cm the light intensity is 64 units. This information is used to answer questions 8 to 10.

8 Which of these is the equation that links L and d?

Ⓐ $L = \frac{40000}{d^2}$ Ⓑ $L = 40000d^2$ Ⓒ $L = \frac{1600}{d^2}$ Ⓓ $L = 1600d^2$ — 1

9 Which of these is closest to the intensity of the light at a distance of 15 cm?

Ⓐ 38 units Ⓑ 64 units Ⓒ 178 units Ⓓ 2667 units — 1

10 At what distance will the light have an intensity of 16 units?

Ⓐ 6.25 cm Ⓑ 40 cm Ⓒ 50 cm Ⓓ 100 cm — 1

In a camp there is enough food to last r refugees t days. On a particular day there are 5000 refugees in the camp with enough food to last 10 days. This information is used to answer questions 11 to 13.

11 Which of these is the equation that links t and r?

Ⓐ $t = 5000r$ Ⓑ $t = \frac{500}{r}$ Ⓒ $t = \frac{50000}{r}$ Ⓓ $t = 500r$ — 1

12 How long will the food last if 8000 refugees are in the camp?

Ⓐ 6.25 days Ⓑ 6.5 days Ⓒ 6.75 days Ⓓ 6.85 days — 1

13 How many **more** refugees (than 5000) could be fed if the food only needs to last 4 days?

Ⓐ 6500 Ⓑ 7000 Ⓒ 7250 Ⓓ 7500 — 1

Rates and Proportion

LEVEL 3 TEST PART B

Write answers and working in the space provided.

Marks

14 Given that $M \propto n^2$, what is the effect on M when the value of n is:

a doubled

b multiplied by 4

c divided by 2

6

15 The power E in kilowatts needed to run a boat varies with the cube of its speed s in m/s. It takes 400 kW to run a boat at 3 m/s.

a Write a formula linking E with v.

b What amount of power is needed to run a boat at 3.6 m/s?

c At what speed is the boat travelling if it takes 1097.6 kW to run?

6

16 In the table $y \propto \frac{1}{x}$.

x	4	B	1.6
y	0.8	0.4	A

a Find the constant of proportionality k.

b Find the value of A.

c Find the value of B.

6

17 The rate r of vibration of a length of string under constant tension is inversely proportional to its length l cm. A 48-cm length of string vibrates 160 times a second.

a Write a formula linking r with l.

b How many times per second will a 60-cm length of string vibrate?

c What is the length of string that vibrates 240 times per second?

6

18 It is known that T varies inversely with the square root of n. If $T = 5$ and $n = 16$, find the value of n when $T = 2.5$.

3

Total marks

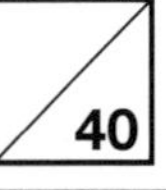

40

Number and Algebra Test: Level 1

STRAND TEST — PART A

Time allowed: 50 minutes

Marks

1 Which of these is 438×18.6321, correct to three significant figures? [1]

Ⓐ 816 Ⓑ 817 Ⓒ 8160 Ⓓ 8161

2 Which of these is the smallest? [1]

Ⓐ 0.016 Ⓑ 15% Ⓒ 6% Ⓓ 0.106

3 Emma's hourly rate of pay is \$23.60. How many hours will she work to be paid \$849.60? [1]

Ⓐ 21 Ⓑ 31 Ⓒ 35 Ⓓ 36

4 Elizabeth receives time-and-a-half pay when she works on Sundays. What will she be paid if she works 6 hours on a Sunday if her pay rate is \$18 per hour? [1]

Ⓐ \$135 Ⓑ \$144 Ⓒ \$162 Ⓓ \$184

5 Expand $5(3 - 2y)$. [1]

Ⓐ $5y$ Ⓑ $13y$ Ⓒ $15 - 2y$ Ⓓ $15 - 10y$

6 Simplify $\frac{3p}{7} + \frac{6p}{7}$. [1]

Ⓐ $\frac{9p}{7}$ Ⓑ $\frac{9p}{14}$ Ⓒ $\frac{18p}{49}$ Ⓓ $\frac{18p^2}{49}$

7 Which of these is the value of $2^4 \times 2^3$? [1]

Ⓐ 2^7 Ⓑ 2^{12} Ⓒ 4^7 Ⓓ 4^{12}

8 Andrew solved an equation correctly. If his answer was $y = -3$, which of these could be his equation? [1]

Ⓐ $3y + 1 = -10$ Ⓑ $4y = 1$ Ⓒ $3 - y = 6$ Ⓓ $2y + 1 = -3$

9 Solve the inequality $4x + 2 > 10$. [1]

Ⓐ $x < 2$ Ⓑ $x > 2$ Ⓒ $x < 3$ Ⓓ $x > 3$

The number plane is used to answer questions 10 and 11.

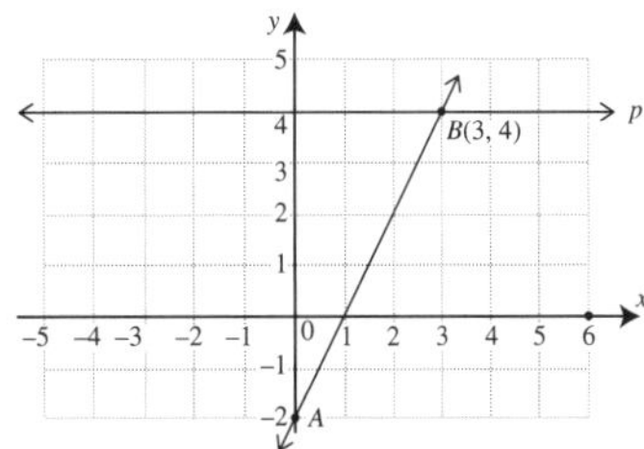

10 Which of these is the equation of the line p? [1]

Ⓐ $x = 3$ Ⓑ $y = 3$ Ⓒ $x = 4$ Ⓓ $y = 4$

11 Which of these is the gradient of the line AB? [1]

Ⓐ $\frac{4}{3}$ Ⓑ $\frac{3}{4}$ Ⓒ $\frac{5}{4}$ Ⓓ 2

12 Solve the equation $(d + 4)(d - 1) = 0$. [1]

Ⓐ $d = -4, -1$ Ⓑ $d = -4, 1$ Ⓒ $d = 4, -1$ Ⓓ $d = 4, 1$

13 If $P = kn$, what is the value of P when $n = 24$ and $k = 0.8$? [1]

Ⓐ 1.92 Ⓑ 6.4 Ⓒ 12.2 Ⓓ 19.2

Number and Algebra Test: Level 1

STRAND TEST — PART B

Write answers and working in the space provided.

Marks

14 Evaluate $\sqrt{12.76 - 3.08}$, correct to two decimal places. ____ [1]

15 Rewrite the following as percentages:

a 12 out of 20 ____

b 30 cents out of \$2 ____ [2]

16 Brett is paid \$76 340 p.a. If he is to receive a pay rise of 2%, what will be his new salary? ____ [2]

17 What is the amount of simple interest on an investment of \$7400 at 6% p.a. for 3 years? ____ [2]

18 Anna receives 4 weeks holiday loading at 17.5% of her normal pay each year. If her normal pay is \$1480 per week, find the:

a amount of holiday loading she will receive ____

b total amount she will receive for the 4 weeks ____ [4]

19 Expand and simplify $3(2x - 1) + 4x$. ____ [2]

20 Simplify:

a $2a^0 - (b^2)^0$ ____

b $p^2q^3 \times p^3q$ ____ [3]

21 Solve the equations:

a $4x + 3 = 2x + 7$ ____

b $\frac{3q - 1}{2} = 7$ ____ [4]

22 Solve the inequality $-5x \geq 15$ and graph your solution on the number line. ____ [2]

23 If $P = 2(l + b)$, what is the value of P when $l = 12$ and $b = 7$? ____ [2]

24 For $y = 4x - 2$, find the:

a gradient ____

b y-intercept ____ [2]

Number and Algebra Test: Level 1

STRAND TEST — PART B

Write answers and working in the space provided.

Marks

25 For $y = 2x - 1$, complete the table and graph:

x	0	1	2
y			

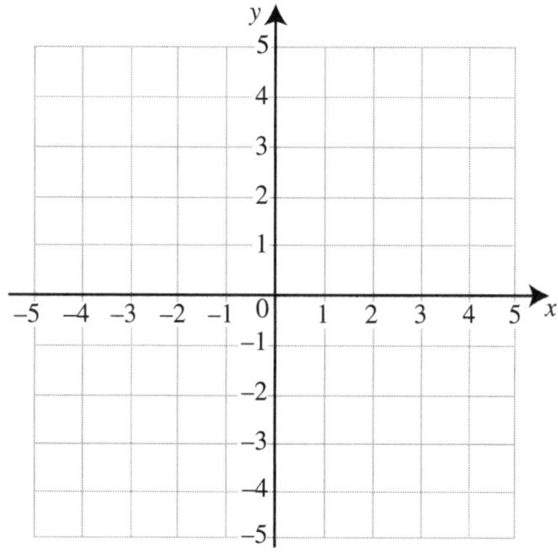

3

26 The points $P(-2, 2)$ and $Q(2, 5)$ are plotted on the number plane.

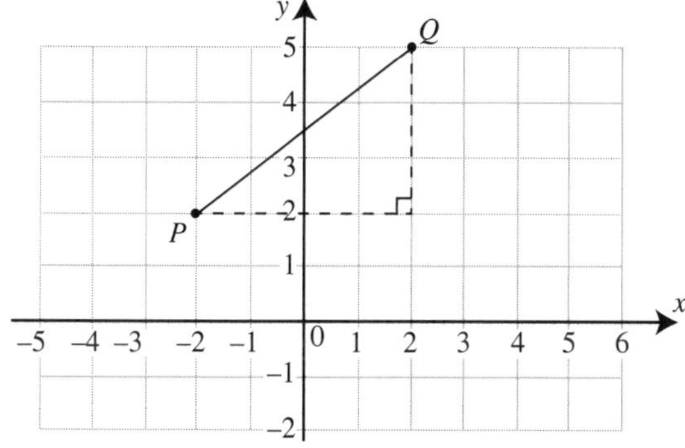

Find the:

a length of PQ

b gradient of PQ

c midpoint of PQ

4

27 Expand $(2a - 1)(a + 3)$.

1

28 Solve $m^2 - 2m - 15 = 0$.

2

29 Using the pronumeral k, write an equation that shows the cost (C) of a hire car is directly proportional to the distance travelled (d).

1

Total marks /50

Number and Algebra Test: Level 2

STRAND TEST — PART A

Time allowed: 50 minutes

Marks

1 Which of these is 580.2917, written in scientific notation, correct to three significant figures?

(A) 5.80×10^2 (B) 5.81×10^2 (C) 5.80×10^6 (D) 5.81×10^6 1

2 Which of these is the best pay?

(A) \$23.80/h normal (B) \$16.50/h time-and-a-half

(C) \$11.85/h double-time (D) \$9.16/h double-time-and-a-half 1

3 What is the simple interest earned on a deposit of \$500 at 4.2% p.a. over 5 years?

(A) \$105 (B) \$112 (C) \$135 (D) \$140 1

4 What is the value of $64^{-\frac{1}{2}}$?

(A) $-\frac{1}{8}$ (B) $\frac{1}{8}$ (C) $\frac{1}{32}$ (D) $63\frac{1}{2}$ 1

5 Which of these is the value of $-12ab + 3ab \times 4$?

(A) 0 (B) 1 (C) $-36ab$ (D) $36ab$ 1

6 Which of these is equal to $8p^4 \div 2p^{-2}$?

(A) $4p^2$ (B) $6p^2$ (C) $4p^6$ (D) $6p^6$ 1

7 Which of these is the solution of the equation $\frac{3p}{2} - 1 = 5$?

(A) $p = 3$ (B) $p = 3\frac{1}{3}$ (C) $p = 3\frac{2}{3}$ (D) $p = 4$ 1

8 Solve the inequality $5(2q - 1) \geq 25$.

(A) $q \geq 2$ (B) $q \geq 2.6$ (C) $q \geq 3$ (D) $q \geq 3.2$ 1

9 Use the formula $A = \frac{h}{2}(a + b)$, to find the value of A if $a = 2.8$, $b = 4.6$ and $h = 2.4$.

(A) 7.64 (B) 7.96 (C) 8.88 (D) 15.28 1

10 Which of these lines is parallel to $y = 3x - 1$ and passes through the origin?

(A) $y = 3x$ (B) $y = -1$ (C) $y = 3x + 2$ (D) $y = 3x - 3$ 1

11 Which of these is the solution of the equation $p^2 + p - 6 = 0$?

(A) $p = -6, 1$ (B) $p = 6, -1$ (C) $p = -3, 2$ (D) $p = 3, -2$ 1

12 The diagram shows the dimensions of a rectangle $ABCD$.

Which of these is an expression for the area, in square centimetres, of the rectangle?

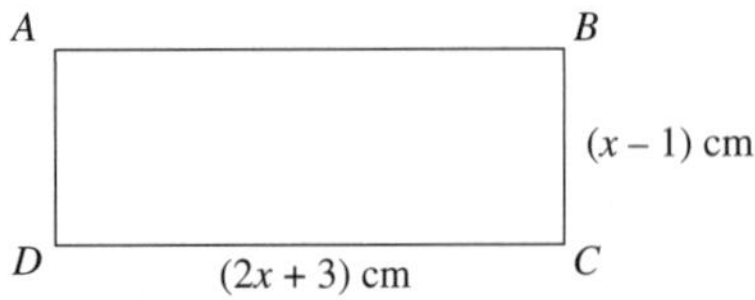

(A) $2x^2 + x - 3$ (B) $2x^2 + 5x - 3$ (C) $2x^2 - 5x + 3$ (D) $2x^2 - x - 3$ 1

13 A ball falls vertically after being dropped. The ball falls a distance d m in a time of t seconds. It is known that d is directly proportional to the square of t. The ball falls 20 m in 2 seconds. Which of these is a formula for d in terms of t?

(A) $d = 5t^2$ (B) $d = 10t^2$ (C) $d = 5\sqrt{t}$ (D) $d = 10\sqrt{t}$ 1

Number and Algebra Test: Level 2

STRAND TEST PART B

Write answers and working in the space provided.

Marks

14 Calculate, leaving your answer correct to two significant figures:

a $\dfrac{12.9+3.87}{2.9-0.054}$ ______ 1

b $\dfrac{1}{\sqrt{3.98+12.8}}$ ______ 1

c $\sqrt{\dfrac{4}{7.6}}-\dfrac{1}{\sqrt{2.3}}$ ______ 1

The tax table is used to answer questions 15 and 16.

Taxable income	Tax on this income
\$0–\$18 200	Nil
\$18 201–\$37 000	19c for each \$1 over \$18 200
\$37 001–\$87 000	\$3572 plus 32.5c for each \$1 over \$37 000
\$87 001–\$180 000	\$19 822 plus 37c for each \$1 over \$87 000
\$180 001 and above	\$54 232 plus 47c for each \$1 over \$180 000

15 Find the tax payable on a taxable income of \$38 490. ______ 2

16 Shannon has a gross income of \$88 520 and tax deductions of \$1360. Find her:

a taxable income ______ b tax payable ______ 2

17 A car can be purchased for a cash price of \$28 000. However, Gareth purchased the car by paying a deposit of 20% and 36 monthly instalments of \$780. How much extra did Gareth pay compared to the cash price of the car?

______ 1

18 Simplify $\dfrac{5ab}{9c}\times\dfrac{3bc}{10a^2}$. ______ 1

19 Expand and simplify $(2b-1)(3b+5)$. ______ 1

20 Simplify:

a $w^5 \div w^{-2}$ ______ b $a^{\frac{2}{3}}\times a^{\frac{1}{3}}$ ______ 2

21 Solve:

a $3(2q-2)-2(q-4)=14$ b $\dfrac{3m+1}{2}+\dfrac{m}{3}=17$ c $3-4(3a-2)\geq 23$

______ 6

Number and Algebra Test: Level 2

STRAND TEST PART B

Write answers and working in the space provided.

Marks

22 If $M = a + bc$, what is the value of b when $M = 16$, $a = 3.2$ and $c = 0.2$? 2

23 Make y the subject in each of the following:

a $2x + 3y - 9 = 0$ **b** $\frac{2x - 1}{3y} = 2$ 4

24 Find the gradient and y-intercept of the line on the number plane.

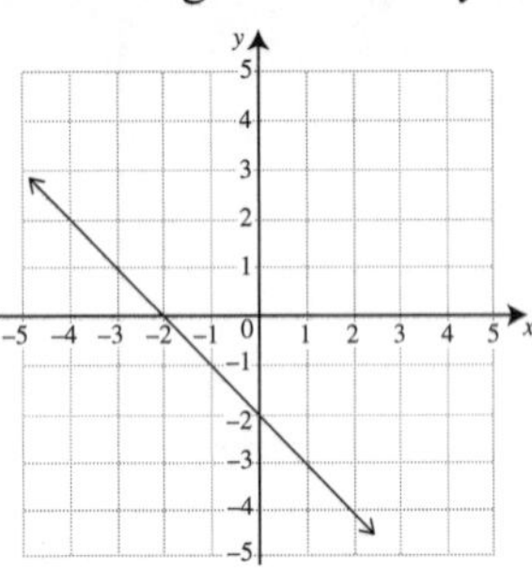

gradient: 1

y-intercept: 1

25 Graph the pair of equations on a number plane and find the point of intersection.

$y = x - 4$ and $y = 2 - x$

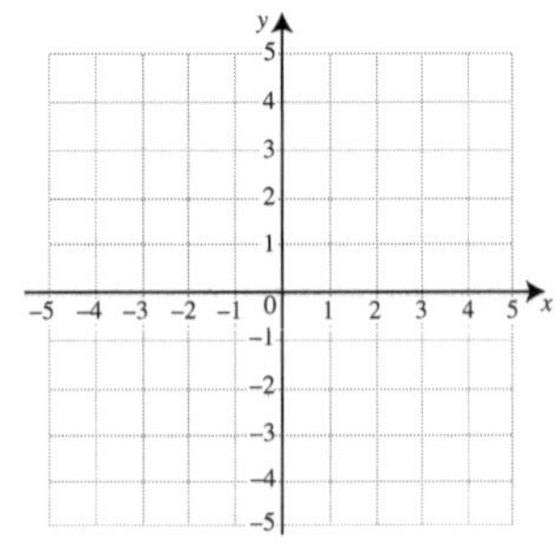

3

26 Factorise $a^2 - 4a - 5ab + 20b$. 2

27 Solve $x^2 - 6x = 7$. 2

28 Simplify $\frac{m^2 - 16}{2m - 8}$. 2

29 M is directly proportional to s^3 and when $s = 4$, $M = 76.8$. By finding the value of k, write a formula linking M with s. 2

Total marks /50

Number and Algebra Test: Level 3

STRAND TEST — PART A

Time allowed: 50 minutes

Marks

1 Which of these is the solution of the inequality $3 - x \geq 1 + x$?

(A) 0 1 2 (B) −2 −1 0 (C) −3 −2 −1 (D) 1 2 3 — 1

2 Which of these is the gradient of the line with equation $4x - 2y + 5 = 0$?

(A) -2 (B) $-\frac{1}{2}$ (C) $\frac{1}{2}$ (D) 2 — 1

3 Which of these is closest to $\dfrac{14.9^3}{\sqrt{13.2 - 0.09}}$?

(A) 65 (B) 252 (C) 910 (D) 914 — 1

4 Simplify $x - \frac{x}{y}$.

(A) $\dfrac{xy - x}{y}$ (B) $xy - x$ (C) $xy - x^2$ (D) $\dfrac{x^2 - x}{y}$ — 1

5 Which of these is the value of m if $5^{2m+1} = 125$?

(A) $m = 1$ (B) $m = 2$ (C) $m = 3$ (D) $m = 5$ — 1

6 Francis invested \$15 000 for 5 years at 6.3% p.a. interest compounded annually. Which of these is closest to the total amount of interest Francis received?

(A) \$1489 (B) \$1576 (C) \$4725 (D) \$5359 — 1

7 Factorise $2a^2 + 5a - 12$.

(A) $(2a - 3)(a + 4)$ (B) $(2a - 1)(a + 12)$ (C) $(2a + 3)(a - 4)$ (D) $2(a - 3)(a + 2)$ — 1

The number plane is used to answer questions 8, 9 and 10.

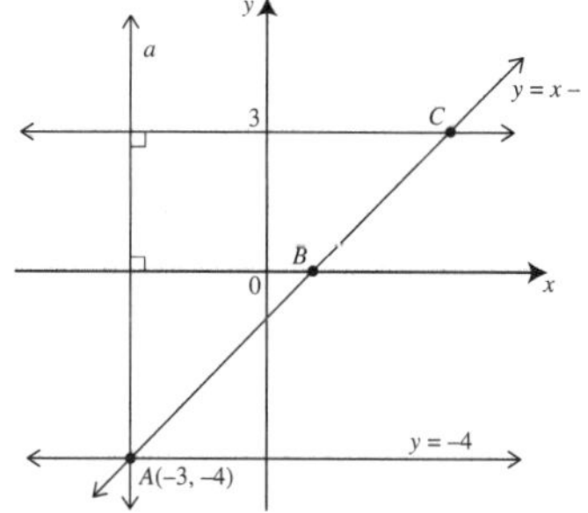

8 Which of these is the equation of the line a?

(A) $x = -3$ (B) $x = -2.5$ (C) $x = -2$ (D) $x = -1.5$ — 1

9 Which of these are the coordinates of the point B?

(A) (0, 1) (B) (1, 0) (C) (2, 0) (D) (0, −1) — 1

10 Which of these are the coordinates of the point C?

(A) (3, 4) (B) (3, 5) (C) (4, 3) (D) (5, 3) — 1

11 Simplify $\dfrac{x-2}{x} \times \dfrac{2x}{x^2 - 4}$.

(A) $\dfrac{1}{x}$ (B) $\dfrac{1}{x+1}$ (C) $\dfrac{2}{x+2}$ (D) $\dfrac{2}{x-1}$ — 1

12 If $3 - p = 5$ and $q = 4p + 5$, which of these is the value of q?

(A) $q = -3$ (B) $q = -1$ (C) $q = 1$ (D) $q = 3$ — 1

13 Which of the following represents an equation where T is inversely proportional to n?

(A) $T = 2n$ (B) $T = \dfrac{\sqrt{n}}{2}$ (C) $T = 2^n$ (D) $T = \dfrac{2}{n}$ — 1

Number and Algebra Test: Level 3

STRAND TEST PART B

Write answers and working in the space provided.

Marks

14 Evaluate, leaving your answer in scientific notation, correct to three significant figures:

a $\dfrac{12.89}{0.03^3}$ 1

b $\sqrt{1.94 \div 276.9}$ 1

15 Express $0.5\dot{2}$ in the form of $\dfrac{p}{q}$ where p and q are integers. 2

16 What amount of interest is earned after investing \$2500 at 4.8% p.a. for 2 years compounded monthly? 2

17 Adam purchases a \$36 000 camper-trailer under the following terms:

- 15% deposit
- balance repaid in monthly instalments of \$952
- 3-year loan.

What annual simple interest rate will be charged? 2

18 Simplify $\dfrac{4c-3}{2} - \dfrac{5-2c}{3}$. 2

19 Expand and simplify $(p-q)^2 - (p+q)^2$.

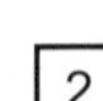

20 Simplify $\left(4a^4b^6\right)^{\frac{3}{2}}$. 1

21 Solve for p: $4^{2p+1} = 8$. 2

Number and Algebra Test: Level 3

STRAND TEST — PART B

Write answers and working in the space provided.

	Marks
22 Solve the following simultaneous equations: $3x - 2y = 16$ and $x + 3y = 9$.	3
23 Solve $\frac{2-y}{5} \geq \frac{3y+2}{3}$.	3
24 For the points $A(-3, 5)$ and $B(1, -3)$, find the: **a** exact length of AB	2
b gradient of AB **c** midpoint of AB	2
25 Find the equation of the line passing through the point (4, 2) and parallel to $y = 1 - 5x$.	2
26 Simplify $\frac{m^2 - m - 6}{m^2 - 4} \div \frac{3m}{m^2 - 2m}$.	3
27 Solve the equations: $y = 2x^2 - 4x + 3$ and $y = x^2 - 5x + 5$.	3
28 It is known that P varies inversely with the square root of x. **a** If $P = 12$, $x = 64$, write an equation linking P and x. **b** Find the value of x when $P = 24$.	4

Total marks /50

Pythagoras' Theorem

PRETEST — LEVEL 1

Time allowed: 10 minutes — **Total marks: 10**

Marks

1 Which of these is the positive solution of the equation $x^2 = 16$?

(A) $x = 4$ (B) $x = 8$ (C) $x = 14$ (D) $x = 256$ — 1

2 What is the value of $2^2 + 3^2$?

(A) 10 (B) 13 (C) 16 (D) 25 — 1

The diagram shows a rectangle and is used to answer questions 3 and 4.

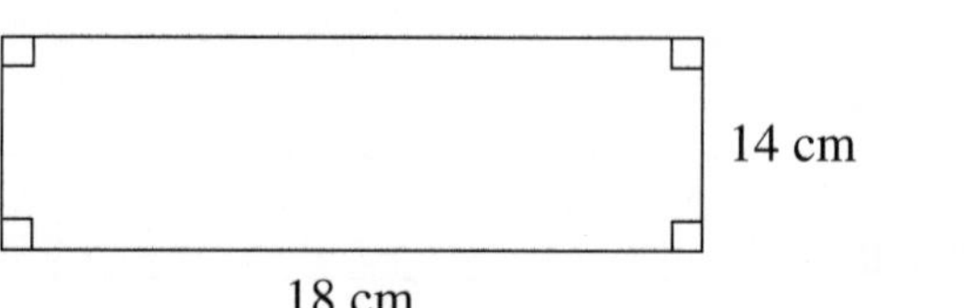

3 Which of these is the area of the rectangle?

(A) 32 cm^2 (B) 64 cm^2 (C) 126 cm^2 (D) 252 cm^2 — 1

4 Which of these is the perimeter of the rectangle?

(A) 32 cm (B) 64 cm (C) 126 cm (D) 252 cm — 1

5 Which of these is a square number?

(A) 12 (B) 28 (C) 48 (D) 144 — 1

6 Which of these is the value of $17^2 - 15^2$?

(A) 8 (B) 16 (C) 64 (D) 169 — 1

The diagram shows a triangle and is used to answer questions 7 and 8.

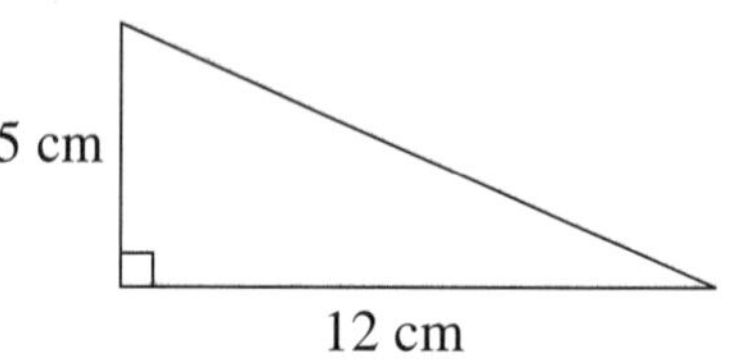

7 What is the area of the triangle?

(A) 13 cm^2 (B) 17 cm^2 (C) 30 cm^2 (D) 60 cm^2 — 1

8 Which of these is the length of the missing side of the triangle?

(A) less than 5 cm (B) between 5 cm and 12 cm

(C) between 12 cm and 17 cm (D) more than 17 cm — 1

9 What is $\sqrt{22}$ correct to one decimal place?

(A) 4.6 (B) 4.7 (C) 5.0 (D) 11.0 — 1

10 Which of these, when written correct to two decimal places, equals 8.37?

(A) $\sqrt{58}$ (B) $\sqrt{67}$ (C) $\sqrt{68}$ (D) $\sqrt{70}$ — 1

Total marks

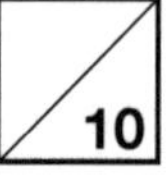

Pythagoras' Theorem

PRETEST

LEVELS 2 AND 3

Time allowed: 10 minutes **Total marks: 10**

Marks

1 Which of these could **not** be the length of the sides of a triangle?

(A) 2 cm, 3 cm, 6 cm (B) 4 cm, 4 cm, 7 cm (C) 5 cm, 5 cm, 5 cm (D) 1 cm, 7 cm, 7 cm 1

2 What is the positive value of x in the equation $x^2 = 3^2 + 4^2$?

(A) $x = 5$ (B) $x = 7$ (C) $x = 14$ (D) $x = 25$ 1

3

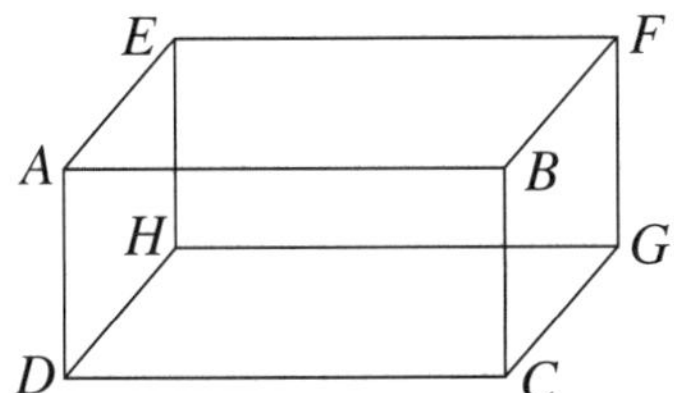

In the rectangular prism, which of these is the longest length?

(A) AB (B) BC (C) AF (D) AG 1

4 Which of these is arranged in ascending order?

(A) $5, 6, \sqrt{33}$ (B) $\sqrt{58}, 8, \sqrt{63}$ (C) $2, 3, \sqrt{12}$ (D) $\sqrt{37}, 6, \sqrt{31}$ 1

5 What is the positive value of x in the equation $x^2 = 10^2 - 6^2$?

(A) $x = 8$ (B) $x = 12$ (C) $x = 16$ (D) $x = 64$ 1

6

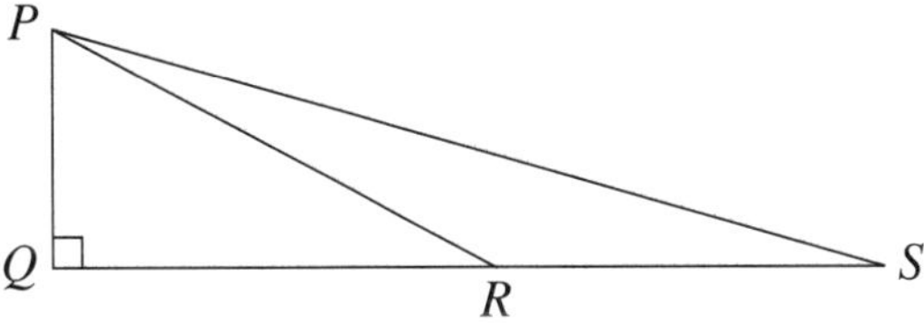

Which of these sides is the longest?

(A) PR (B) PS (C) PQ (D) QS 1

7 If $x^2 - 60 = 225$, which of these is closest to the positive value of x?

(A) $x = 12.3$ (B) $x = 12.7$ (C) $x = 17$ (D) $x = 20$ 1

8 Between which two consecutive even numbers does $\sqrt{240}$ lie?

(A) 15 and 16 (B) 14 and 16 (C) 16 and 18 (D) 14 and 18 1

9 Which of these is a value of b for the equation $b^2 = 5^2 + 6^2$?

(A) $b = 7$ (B) $b = 11$ (C) $b = \sqrt{61}$ (D) $b = 121$ 1

10 A rectangle has a width of 4 cm and an area of 24 cm^2. Which of the following is true about the length of its diagonal?

(A) less than 4 cm (B) between 4 cm and 6 cm

(C) between 6 cm and 10 cm (D) more than 10 cm 1

Total marks /10

Pythagoras' Theorem

LEVEL 1 TEST — PART A

Time allowed: 40 minutes

Marks

The diagrams show triangles ABC, PQR and XYZ and are used to answer questions 1 to 5.

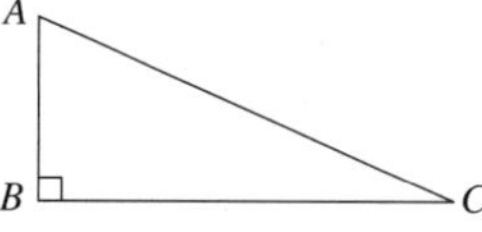

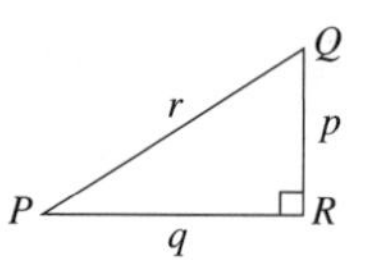

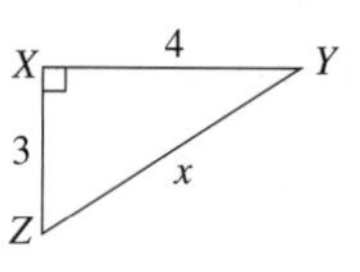

1 In triangle ABC, which side is the hypotenuse?

Ⓐ AC Ⓑ BC Ⓒ AB Ⓓ none of these — 1

2 Using triangle ABC, which of the following is correct?

Ⓐ $AB^2 = BC^2 + AC^2$ Ⓑ $AC^2 = AB^2 + BC^2$ Ⓒ $BC^2 = AC^2 + AB^2$ Ⓓ $BC^2 = AC^2 \times AB^2$ — 1

3 In triangle PQR, which side is the hypotenuse?

Ⓐ p Ⓑ q Ⓒ r Ⓓ none of these — 1

4 Using triangle PQR, which of the following is correct?

Ⓐ $r^2 = p^2 + q^2$ Ⓑ $r^2 = p^2 - q^2$ Ⓒ $q^2 = p^2 + r^2$ Ⓓ $p^2 = r^2 + q^2$ — 1

5 Using triangle XYZ, which of these shows the correct relationship?

Ⓐ $x = 3 + 4$ Ⓑ $x^2 = 4^2 - 3^2$ Ⓒ $x^2 = 3^2 + 4^2$ Ⓓ $x = 4 - 3$ — 1

6 Which of these are sides of a right-angled triangle?

Ⓐ 6 cm, 7 cm, 8 cm Ⓑ 8 cm, 10 cm, 12 cm Ⓒ 12 cm, 16 cm, 20 cm Ⓓ 15 cm, 20 cm, 22 cm — 1

7 If $x^2 = 10^2 + 24^2$, what is the value of x^2?

Ⓐ 576 Ⓑ 676 Ⓒ 776 Ⓓ 1156 — 1

8 Which of these is a Pythagorean triad?

Ⓐ {3, 12, 13} Ⓑ {3, 4, 6} Ⓒ {1, 2, 3} Ⓓ {6, 8, 10} — 1

9 If $c^2 = a^2 - b^2$ and $a = 9$, $b = 4$, which of these is the value of c^2?

Ⓐ 25 Ⓑ 35 Ⓒ 45 Ⓓ 65 — 1

The diagrams show triangles WXY, TUV, FGH and JKL and are used to answer questions 10 to 13.

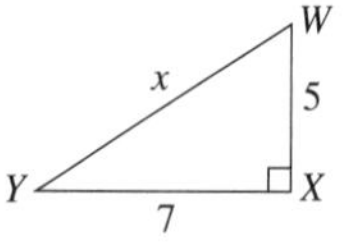

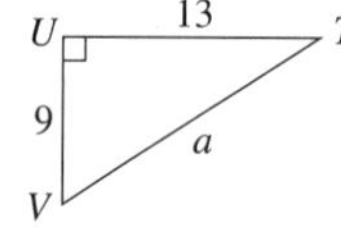

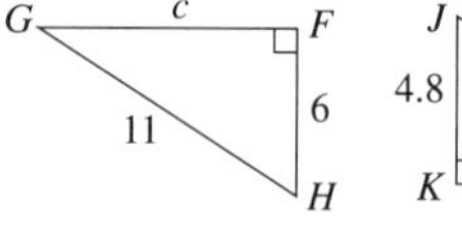

10 Using triangle WXY, which of these shows the correct relationship?

Ⓐ $x^2 = 7^2 - 5^2$ Ⓑ $x = 7 - 5$ Ⓒ $x^2 = 5^2 + 7^2$ Ⓓ $x^2 = 5^2 \times 7^2$ — 1

11 Using triangle TUV, which of these shows the correct relationship?

Ⓐ $a = 13^2 + 9^2$ Ⓑ $a^2 = 13^2 + 9^2$ Ⓒ $a^2 = 13^2 - 9^2$ Ⓓ $a^2 = 9^2 - 13^2$ — 1

12 Using triangle FGH, which of these shows the correct relationship?

Ⓐ $c^2 = 11^2 + 6^2$ Ⓑ $c^2 = 11 + 6$ Ⓒ $c^2 = 6^2 - 11^2$ Ⓓ $c^2 = 11^2 - 6^2$ — 1

13 Using triangle JKL, which of these shows the correct relationship?

Ⓐ $m^2 = 7.2^2 - 4.8^2$ Ⓑ $m^2 = 4.8^2 - 7.2^2$ Ⓒ $m^2 = 4.8^2 + 7.2^2$ Ⓓ $m = 7.2 - 4.8$ — 1

Pythagoras' Theorem

LEVEL 1 TEST — PART B

Write answers and working in the space provided.

Marks

14 Evaluate the following, correct to two decimal places:

a $\sqrt{9+36}$ ______ 1

b $\sqrt{4+16}$ ______ 1

c $\sqrt{49+81}$ ______ 1

15 Use Pythagoras' Theorem to find the value of the pronumeral:

a
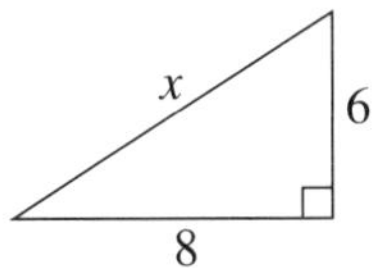

b
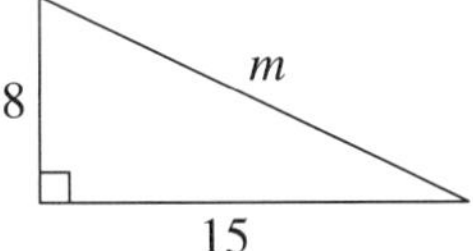

c
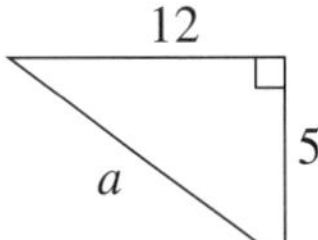

6

d
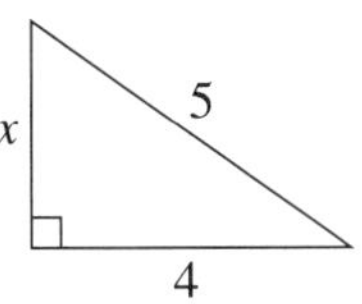

e
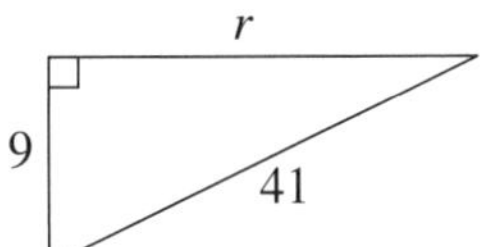

f
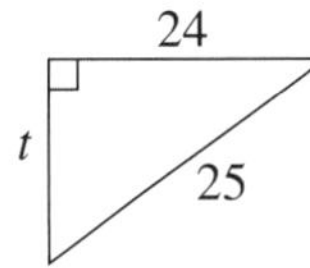

6

16 Find the value of the pronumeral, correct to two decimal places:

a
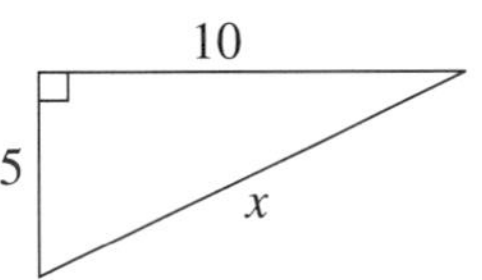

b
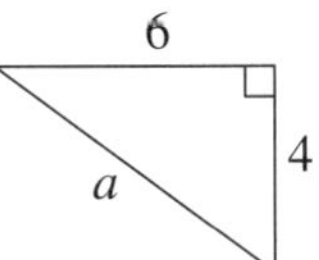

c
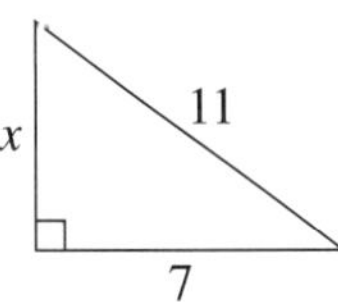

6

d
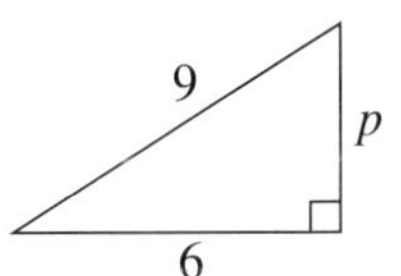

e
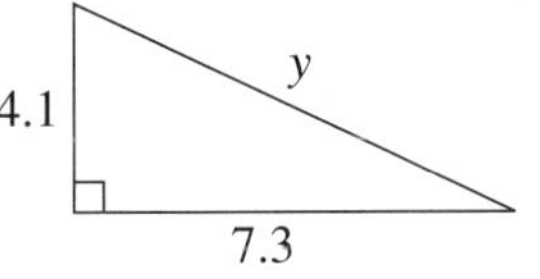

f
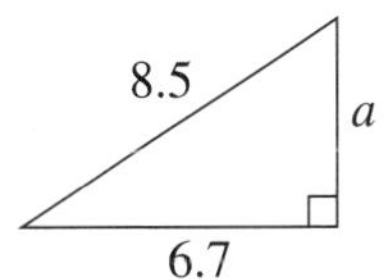

6

Total marks /40

Pythagoras' Theorem

LEVEL 2 TEST

PART A

Time allowed: 40 minutes

Marks

The diagrams show triangles ABC, FGH, LMN and PQR and are used to answer questions 1 to 4.

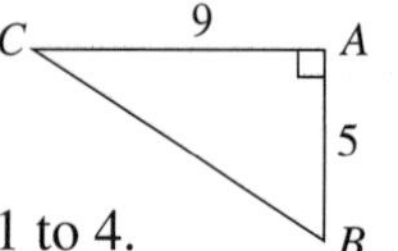

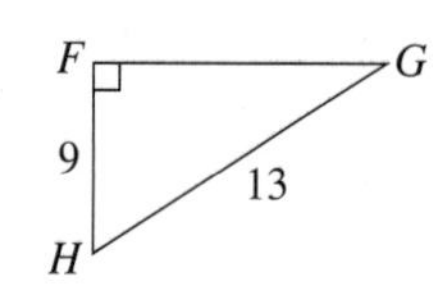

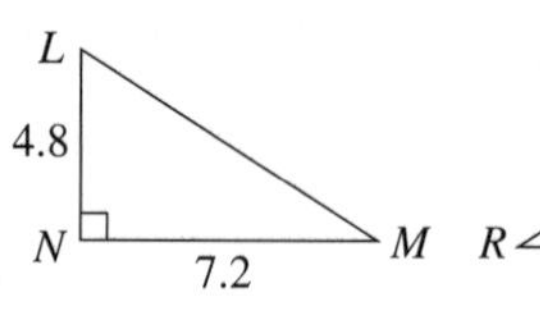

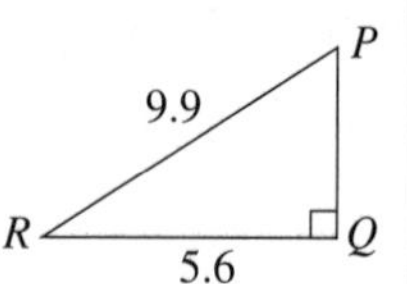

1 Using triangle ABC, which of these is closest to the length of BC?

Ⓐ 7.48 Ⓑ 7.49 Ⓒ 10.29 Ⓓ 10.30 1

2 Using triangle FGH, which of these is closest to the length of FG?

Ⓐ 9.37 Ⓑ 9.38 Ⓒ 15.81 Ⓓ 15.82 1

3 Using triangle LMN, which of these is closest to the length of LM?

Ⓐ 5.36 Ⓑ 5.37 Ⓒ 8.64 Ⓓ 8.65 1

4 Using triangle PQR, which of these is closest to the length of PQ?

Ⓐ 8.16 Ⓑ 8.17 Ⓒ 11.37 Ⓓ 11.38 1

5 A right triangle has shorter sides of 8 cm and 12 cm. Which of these is the exact length of the hypotenuse?

Ⓐ $\sqrt{40}$ cm Ⓑ $\sqrt{148}$ cm Ⓒ $\sqrt{192}$ cm Ⓓ $\sqrt{208}$ cm 1

6 Which of these is **not** a Pythagorean triad?

Ⓐ {16, 36, 39} Ⓑ {11, 60, 61} Ⓒ {14, 48, 50} Ⓓ {13, 84, 85} 1

7 How far up a wall will a 2.4-m ladder reach if the foot of the ladder is placed 1 m from the wall? Give your answer correct to one decimal place.

Ⓐ 2.6 m Ⓑ 2.2 m Ⓒ 1.9 m Ⓓ 1.8 m 1

8 What is the length of a diagonal of a rectangle with sides 65 cm and 72 cm?

Ⓐ 90 cm Ⓑ 91 cm Ⓒ 93 cm Ⓓ 97 cm 1

9 Which of these is the hypotenuse of an isosceles right triangle of side 2 cm?

Ⓐ 2 cm Ⓑ $\sqrt{6}$ cm Ⓒ $\sqrt{8}$ cm Ⓓ 6 cm 1

The diagrams are used to answer questions 10 to 13.

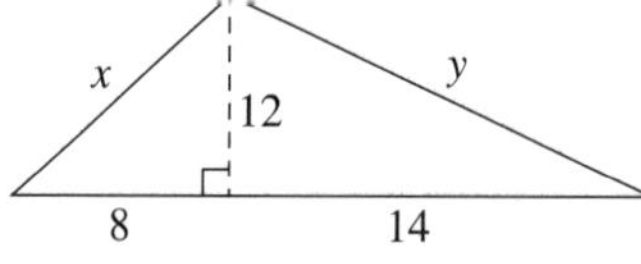

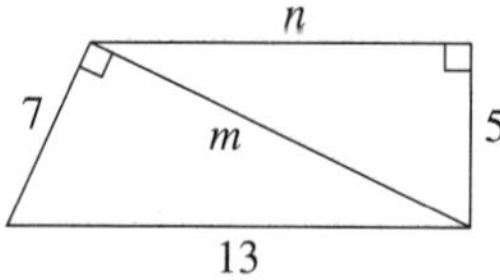

10 Which of these is the value of x, correct to two decimal places?

Ⓐ 8.94 Ⓑ 8.95 Ⓒ 14.42 Ⓓ 14.43 1

11 Which of these is the value of y, correct to three significant figures?

Ⓐ 7.21 Ⓑ 7.22 Ⓒ 18.4 Ⓓ 18.5 1

12 Which of these is the value of m, leaving your answer as a surd?

Ⓐ $\sqrt{36}$ Ⓑ $\sqrt{96}$ Ⓒ $\sqrt{106}$ Ⓓ $\sqrt{120}$ 1

13 Which of these is the value of n, correct to two significant figures?

Ⓐ 9.6 Ⓑ 9.7 Ⓒ 9.8 Ⓓ 9.9 1

Pythagoras' Theorem

LEVEL 2 TEST — PART B

Write answers and working in the space provided.

Marks

14 Find the value of the pronumeral, leaving your answer as a surd:

a

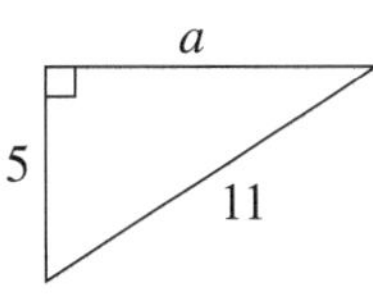

b

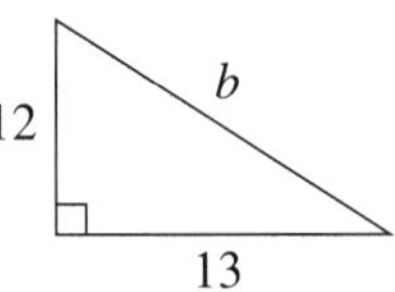

c

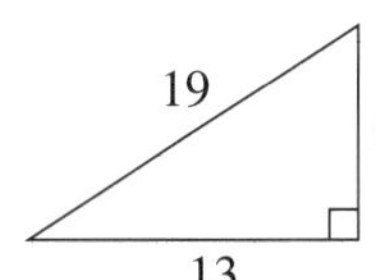

6

15 What is the value of the pronumeral, giving your answer to two significant figures?

a

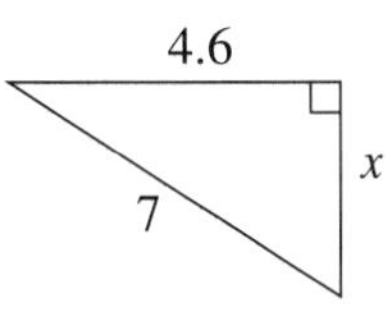

b

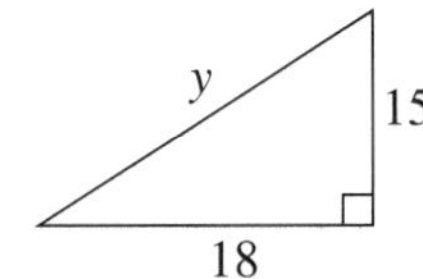

c

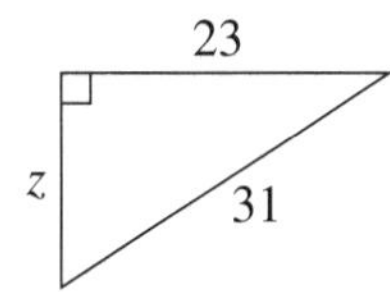

6

16 Find the values of the pronumerals, correct to three decimal places:

a

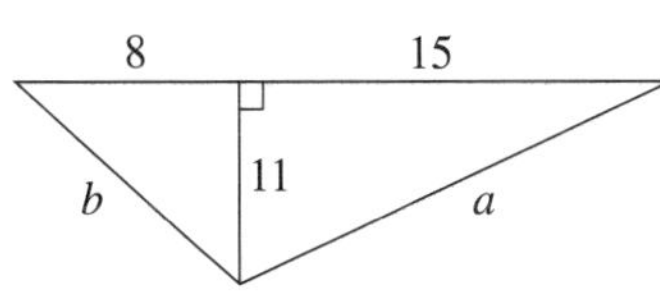

b

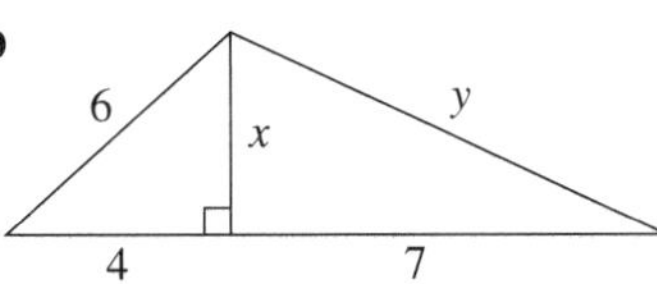

c

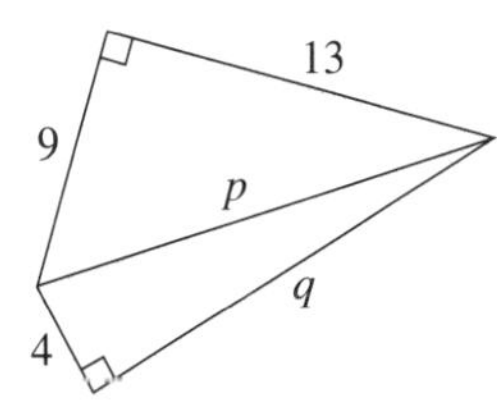

12

17 A cyclist leaves A and rides 18 km east to B before turning south and riding 24 km to C. What is the straight-line distance from A to C?

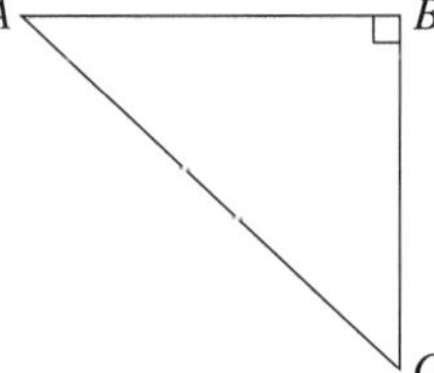

3

Total marks /40

Pythagoras' Theorem

LEVEL 3 TEST — PART A

Time allowed: 40 minutes

Marks

1 The longest side of a right-angled triangle is 15 cm. If the shortest side is 9 cm, how long is the other side?

(A) 6 cm (B) 12 cm (C) 12.5 cm (D) 13 cm — 1

2 The hypotenuse of an isosceles right triangle is 4 cm. What is the area?

(A) 4 cm^2 (B) 8 cm^2 (C) 16 cm^2 (D) 32 cm^2 — 1

The diagram is used to answer questions 3 to 5.

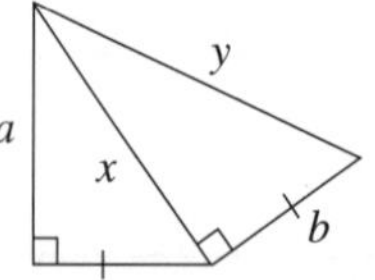

3 If $a = 4$ and $b = 3$, what is the value of x?

(A) 4 (B) 5 (C) $\sqrt{32}$ (D) 6 — 1

4 If $a = 4$ and $b = 3$, what is the value of y?

(A) 5 (B) $\sqrt{34}$ (C) 6 (D) $\sqrt{41}$ — 1

5 Which of these is an expression for y^2 in terms of a and b?

(A) $y^2 = a^2 + b^2$ (B) $y^2 = a^2 + 2b^2$ (C) $y^2 = 2a^2 + b^2$ (D) $y^2 = 2a^2 + 2b^2$ — 1

6 The perimeter of a right-angled triangle is 12 units. Which of these could be its area?

(A) 6 units^2 (B) 8 units^2 (C) 10 units^2 (D) 12 units^2 — 1

7 A rectangle has its sides in the ratio 2 : 1 and the length of its diagonal is 10 cm. Which of these is the area of the rectangle?

(A) 12 cm^2 (B) 20 cm^2 (C) 40 cm^2 (D) 48 cm^2 — 1

8 What is the exact length of the hypotenuse in a right triangle with shorter sides of $\sqrt{11}$ cm and $\sqrt{5}$ cm?

(A) 4 cm (B) 8 cm (C) 12 cm (D) 16 cm — 1

9 Which of these is the perimeter of a rhombus with diagonals 12 cm and 16 cm?

(A) 32 cm (B) 36 cm (C) 40 cm (D) 48 cm — 1

10 A yacht leaves a buoy and sails 3 km south and then 2 km east. Which of these is the distance, to the nearest metre, of the yacht from the buoy?

(A) 2236 m (B) 2237 m (C) 3605 m (D) 3606 m — 1

11 A rectangular box has dimensions 5 cm by 4 cm by 3 cm. Which of these is the length of the longest toothpick that can fit inside the box?

(A) $\sqrt{50}$ cm (B) $\sqrt{60}$ cm (C) $\sqrt{80}$ cm (D) 10 cm — 1

12 The altitude of an equilateral triangle is $\sqrt{27}$ cm. What is the perimeter of the triangle?

(A) $\sqrt{3}$ cm (B) 3 cm (C) 9 cm (D) 18 cm — 1

13 Two trees are 24 m apart and are 16 m and 26 m tall. Which of these is the distance between the tops of the trees?

(A) 26 cm (B) 30 cm (C) 34 cm (D) 36 cm — 1

Pythagoras' Theorem

LEVEL 3 TEST — PART B

Write answers and working in the space provided.

Marks

The diagrams show three solids and are used to answer questions 14 to 16. All units are in centimetres.

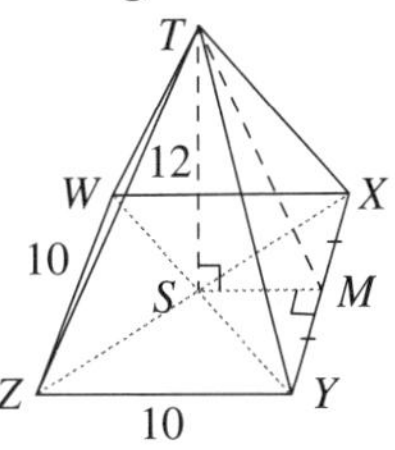

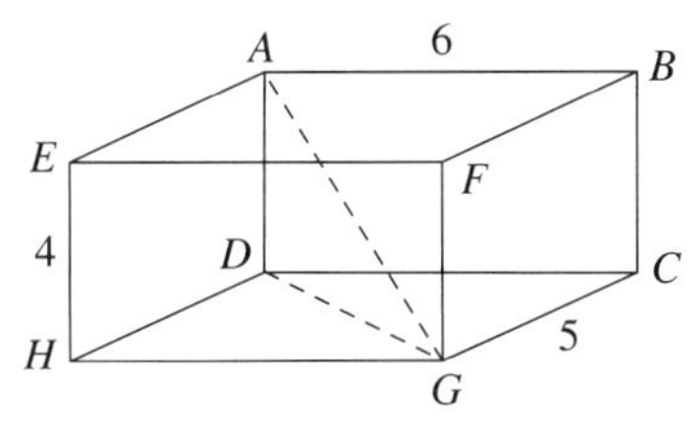

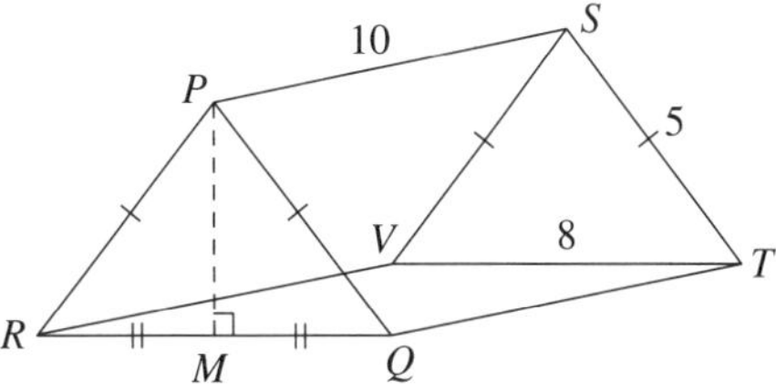

14 Using the square-based pyramid $WXYZT$, find:

a the length of SM

b the length of TM

c the area of triangle TXY

6

15 Using the rectangular prism $ABCDEFG$, find, correct to two decimal places:

a the length of DG

b the area of triangle ADG

c the length of AG

6

16 Using the triangular prism $PQRSTV$, find the exact:

a height of PM

b length of MT

c length of PT

6

17 Using the cone PRQ, find the:

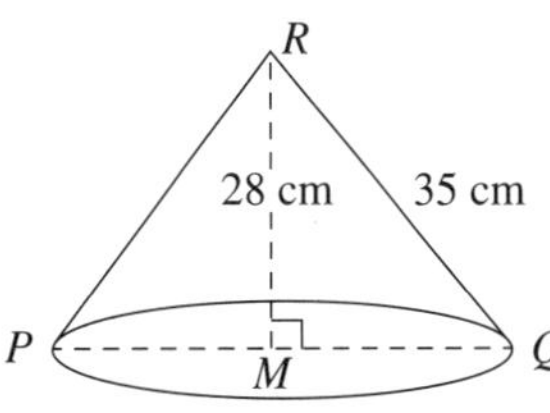

a length of PQ

b area of the circular base, in terms of π

4

18

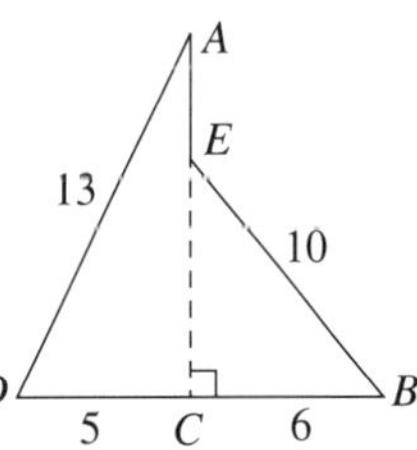

a Find the length of AE.

b Find the area of $AEBD$.

5

Total marks

Area, Surface Area and Volume

PRETEST — LEVEL 1

Time allowed: 10 minutes **Total marks: 10**

Marks

1 Which of these is the formula used to find the area of a circle?

(A) $A = 2\pi r$ (B) $A = 2\pi r^2$ (C) $A = \pi r^2$ (D) $A = \pi d^2$ 1

2 Which of these is the area of a square with side length 1.6 cm?

(A) 2.56 cm^2 (B) 3.2 cm^2 (C) 6.4 cm^2 (D) 10.24 cm^2 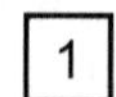

A circle has a radius of 8 cm. This information is used to answer questions 3 and 4.

3 What is the circumference of the circle, correct to two decimal places?

(A) 25.13 cm (B) 25.14 cm (C) 50.26 cm (D) 50.27 cm 1

4 Which of these is the area of the circle, correct to three decimal places?

(A) 194.348 cm^2 (B) 194.349 cm^2 (C) 201.062 cm^2 (D) 201.063 cm^2 

5 What is the volume of a cube with side length 12 cm?

(A) 36 cm^3 (B) 48 cm^2 (C) 144 cm^3 (D) 1728 cm^3 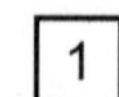

6

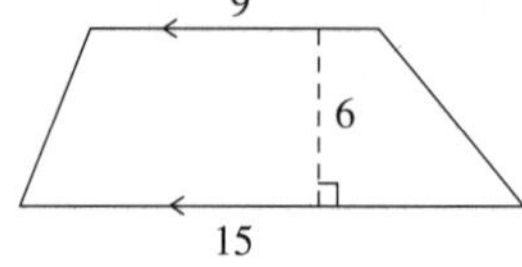

Which of these is the area of the trapezium?

(A) 30 cm^2 (B) 72 cm^2 (C) 405 cm^2 (D) 810 cm^2 1

7 A parallelogram has a base of 16 cm and a height of 12 cm. What is the area?

(A) 52 cm^2 (B) 96 cm^2 (C) 192 cm^2 (D) 1885 cm^2 1

The diagrams show three solids and are used to answer questions 8, 9 and 10.

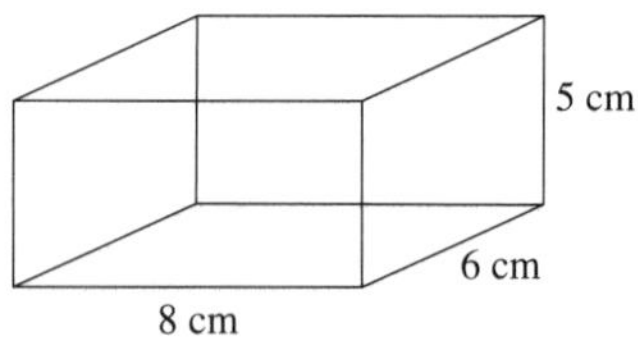

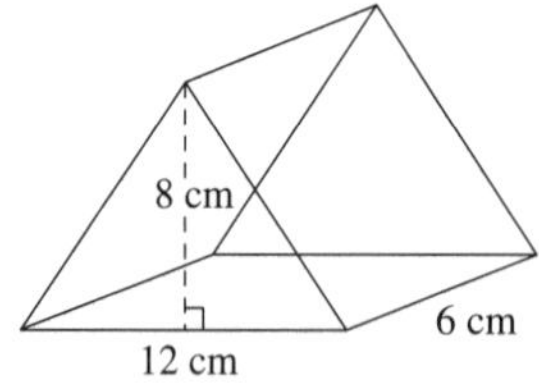

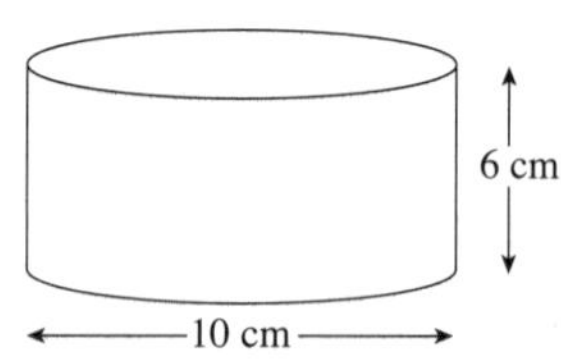

8 Which of these is the volume of the rectangular prism?

(A) 19 cm^3 (B) 53 cm^2 (C) 236 cm^3 (D) 240 cm^3 1

9 Which of these is the volume of the triangular prism?

(A) 26 cm^3 (B) 144 cm^2 (C) 288 cm^3 (D) 576 cm^3 1

10 Which of these is closest to the volume of the cylinder?

(A) 471 cm^3 (B) 600 cm^2 (C) 1131 cm^3 (D) 1885 cm^3 1

Total marks 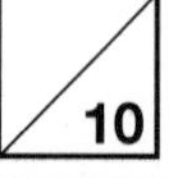

Area, Surface Area and Volume

PRETEST

LEVELS 2 AND 3

Time allowed: 10 minutes **Total marks: 10**

Marks

The diagrams show three shapes and are used to answer questions 1 to 6.

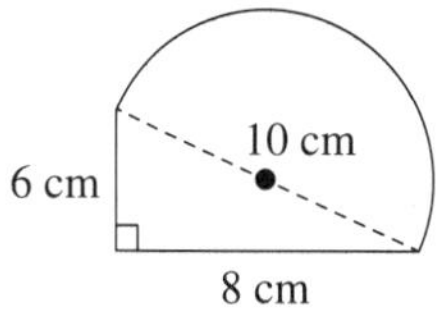

Figure 1

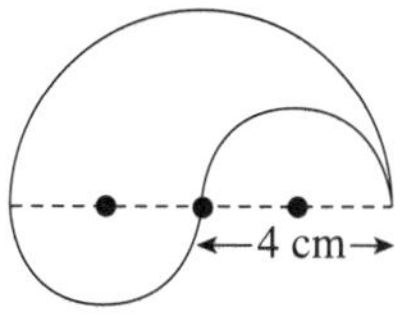

Figure 2

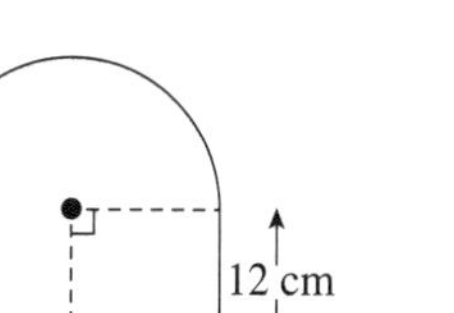

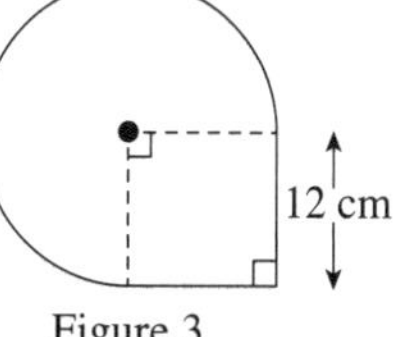

Figure 3

1 Which of these is closest to the perimeter of figure 1?

(A) 21.9 cm (B) 29.7 cm (C) 55.4 cm (D) 76.8 cm [1]

2 Which of these is closest to the area of figure 1?

(A) 63.3 cm^2 (B) 87.3 cm^2 (C) 181.1 cm^2 (D) 205.1 cm^2 [1]

3 Which of these is closest to the perimeter of figure 2?

(A) 18.8 cm (B) 25.1 cm (C) 31.4 cm (D) 37.7 cm [1]

4 Which of these is closest to the area of figure 2?

(A) 25.1 cm^2 (B) 31.4 cm^2 (C) 37.7 cm^2 (D) 100.5 cm^2 [1]

5 Which of these is closest to the perimeter of figure 3?

(A) 75.4 cm (B) 80.5 cm (C) 87.4 cm (D) 99.4 cm [1]

6 Which of these is closest to the area of figure 3?

(A) 153.4 cm^2 (B) 448.8 cm^2 (C) 452.4 cm^2 (D) 483.3 cm^2 [1]

The diagrams show four shapes and are used to answer questions 7 to 10.

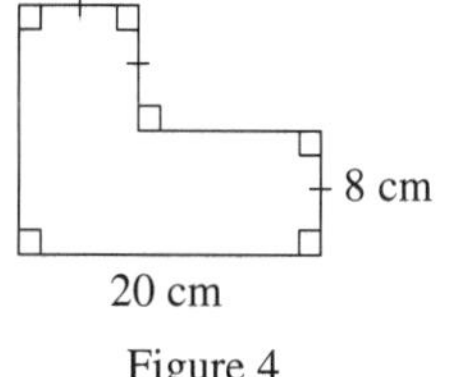

Figure 4

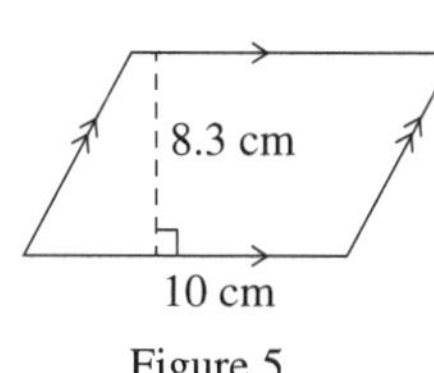

Figure 5

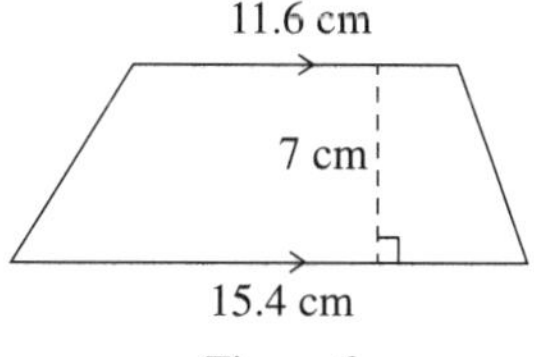

Figure 6

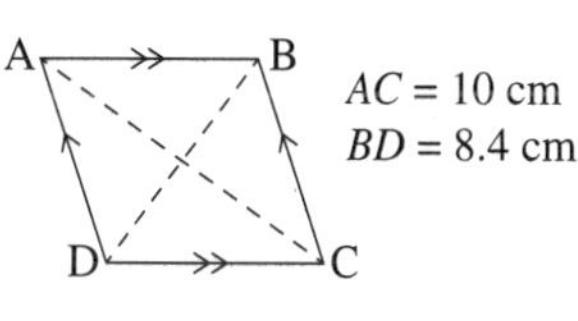

Figure 7

7 Which of these is the area of figure 4?

(A) 176 cm^2 (B) 224 cm^2 (C) 336 cm^2 (D) 464 cm^2 [1]

8 Which of these is the area of figure 5?

(A) 36.6 cm^2 (B) 41.5 cm^2 (C) 80.3 cm^2 (D) 83 cm^2 [1]

9 Which of these is the area of figure 6?

(A) 34 cm^2 (B) 94.5 cm^2 (C) 189 cm^2 (D) 1250.48 cm^2 [1]

10 Which of these is the area of figure 7?

(A) 21 cm^2 (B) 36.8 cm^2 (C) 42 cm^2 (D) 84 cm^2 [1]

Total marks /10

Area, Surface Area and Volume

LEVEL 1 TEST — PART A

Time allowed: 40 minutes

Marks

The diagrams are used to answer questions 1 to 4.

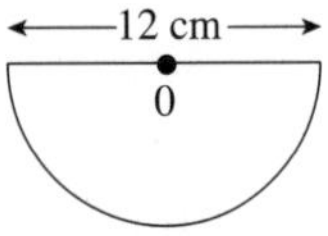

Figure 1 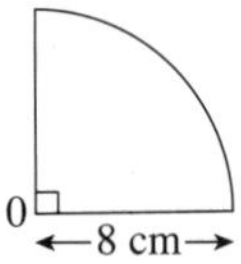 Figure 2 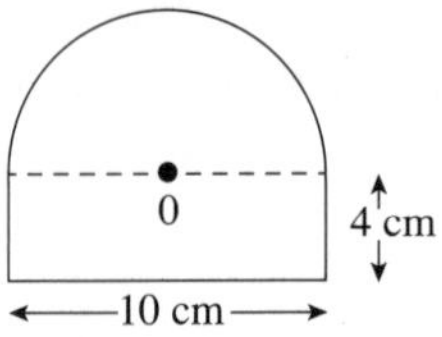 Figure 3 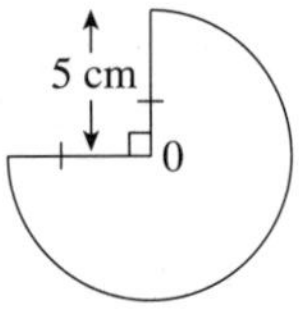 Figure 4

1 Which of these is the area of figure 1, correct to two decimal places?

(A) 56.55 cm² (B) 75.40 cm² (C) 226.19 cm² (D) 452 .39 cm² | 1

2 Which of these is closest to the area of figure 2?

(A) 12.6 cm² (B) 50.3 cm² (C) 100.5 cm² (D) 201.1 cm² | 1

3 What is the area of figure 3, to the nearest square centimetre?

(A) 79 cm² (B) 119 cm² (C) 197 cm² (D) 354 cm² | 1

4 Which of these is the area of figure 4, correct to one decimal place?

(A) 58.9 cm² (B) 68.9 cm² (C) 78 cm² (D) 236 cm² | 1

5 How many rectangular faces are on a triangular prism?

(A) 3 (B) 4 (C) 5 (D) 6 | 1

6 Which of these is the capacity of a cube of side 10 cm?

(A) 30 mL (B) 60 mL (C) 100 mL (D) 1 L | 1

7 What is the surface area of a rectangular prism with dimensions 5 cm, 4 cm, and 3 cm?

(A) 12 cm² (B) 50 cm² (C) 60 cm² (D) 94 cm² | 1

8 Which of these is closest to the volume of a cylinder with radius 8 cm and height 9 cm?

(A) 226 cm³ (B) 452 cm³ (C) 1810 cm³ (D) 2036 cm³ | 1

9 The diagonals of a rhombus are 12 cm and 10 cm. What is the area of the rhombus?

(A) 58.9 cm² (B) 68.9 cm² (C) 60 cm² (D) 120 cm² | 1

The diagrams are used to answer questions 10 to 13.

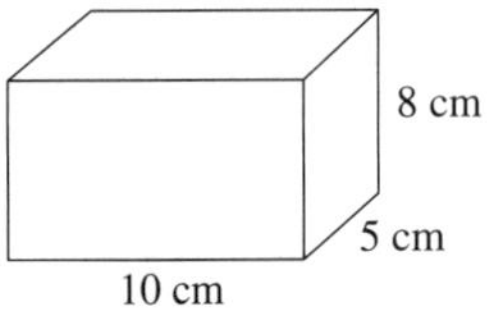

Figure 5 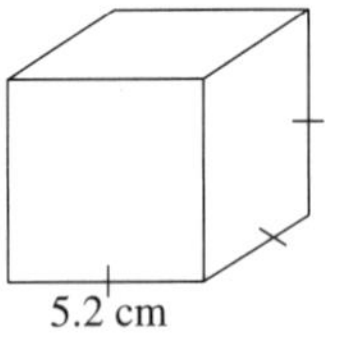 Figure 6 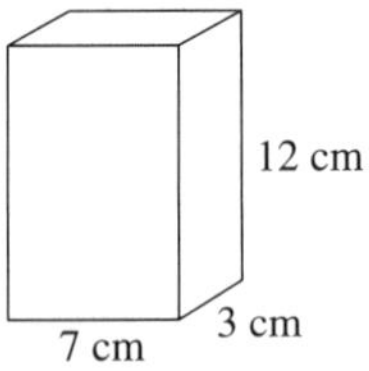 Figure 7 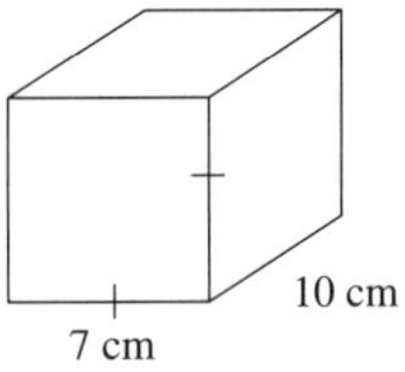 Figure 8

10 Which of these is the surface area of the rectangular prism in figure 5?

(A) 170 cm² (B) 180 cm² (C) 340 cm² (D) 400 cm² | 1

11 What is the surface area of the cube in figure 6?

(A) 27.04 cm² (B) 31.2 cm² (C) 81.12 cm² (D) 162.24 cm² | 1

12 Which of these is the surface area of the rectangular prism in figure 7?

(A) 141 cm² (B) 252 cm² (C) 282 cm² (D) 296 cm² | 1

13 Which of these is the surface area of the square-based prism in figure 8?

(A) 238 cm² (B) 329 cm² (C) 378 cm² (D) 490 cm² | 1

Area, Surface Area and Volume

LEVEL 1 TEST — PART B

Write answers and working in the space provided.

Marks

The diagrams show three solids which are used to answer questions 14 to 16.

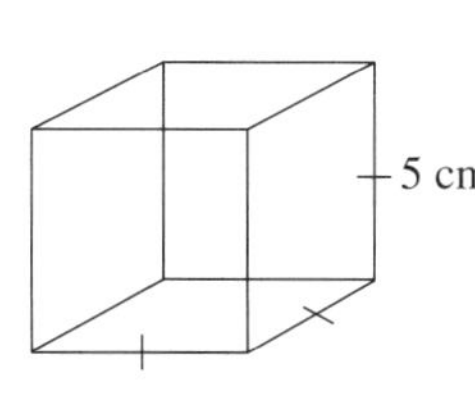

Figure 1

Figure 2

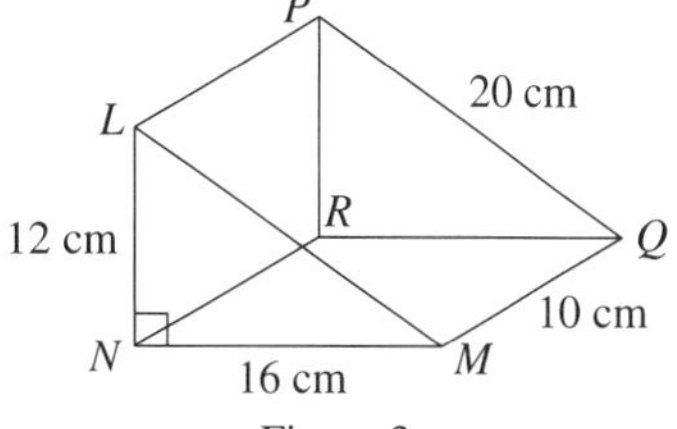

Figure 3

14 Using the cube in figure 1, find the:

a volume

b area of one face

c surface area

6

15 Using the rectangular prism in figure 2, find the:

a volume

b area of the face *BFGC*

c surface area

6

16 Using the triangular prism in figure 3, find the:

a volume

b area of the face *PQML*

c surface area

6

17 Find the capacity, expressed in litres, of the following solids:

a

b

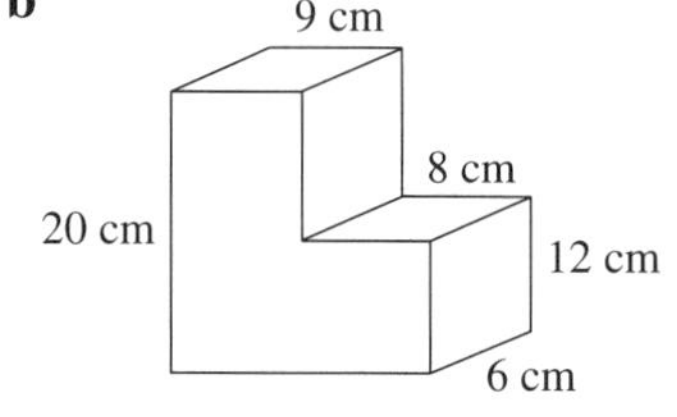

c

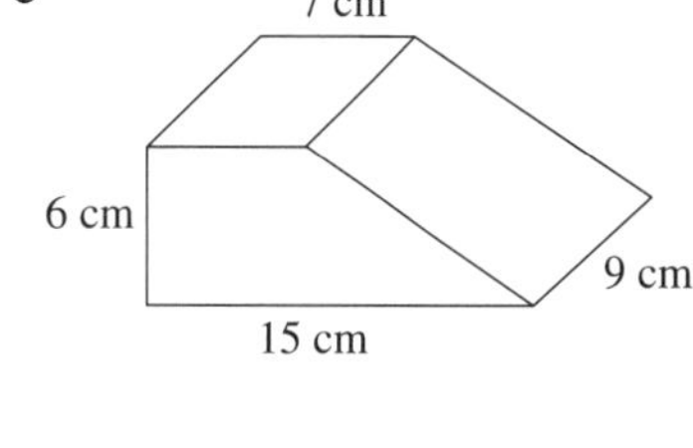

9

Total marks

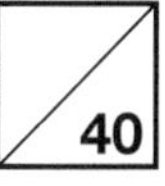

Area, Surface Area and Volume

LEVEL 2 TEST — PART A

Time allowed: 40 minutes

Marks

1 Which of these is equal to 1 m^2?

(A) 10 cm^2 (B) 100 cm^2 (C) 1000 cm^2 (D) 10 000 cm^2 — 1

2 What is the surface area of a rectangular prism with dimensions 8 cm, 6 cm and 4 cm?

(A) 104 cm^2 (B) 164 cm^2 (C) 192 cm^2 (D) 208 cm^2 — 1

3 The surface area of a cube is 24 cm^2. What is the volume of the cube?

(A) 4 cm^3 (B) 8 cm^3 (C) 24 cm^3 (D) 64 cm^3 — 1

4 Which of these is closest to the curved surface area of a cylinder of radius 6 cm and height 9 cm?

(A) 245 cm^2 (B) 246 cm^2 (C) 339 cm^2 (D) 340 cm^2 — 1

5 A square with side x cm is cut out of a rectangle with dimensions $4x$ cm and $3x$ cm. What is the area of the remaining shape?

(A) $11x$ cm^2 (B) $11x^2$ cm^2 (C) 11 cm^2 (D) $(12x^2 - x)$ cm^2 — 1

6 The curved surface area of a cylinder is 20π cm^2. If the height of the cylinder is 5 cm, which of these is the cylinder's total surface area?

(A) 28π cm^2 (B) 32π cm^2 (C) 36π cm^2 (D) 48π cm^2 — 1

The diagrams show four solids and are used to answer questions 7 to 13.

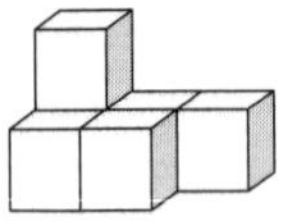

Figure 1 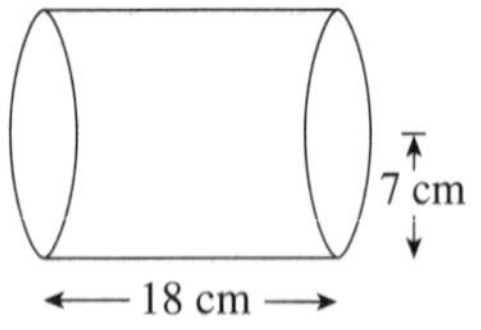 Figure 2 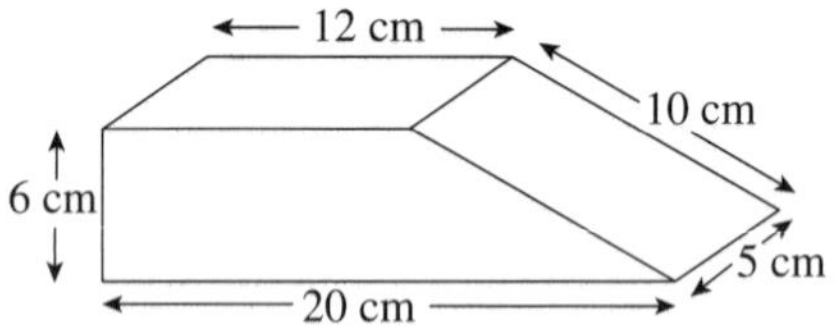 Figure 3 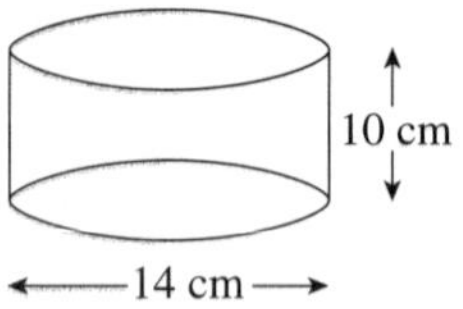 Figure 4

7 In figure 1, six cubes with sides of 1 unit are glued together. What is the surface area of the solid?

(A) 24 $units^2$ (B) 30 $units^2$ (C) 33 $units^2$ (D) 36 $units^2$ — 1

8 Which of these is closest to the volume of the cylinder in figure 2?

(A) 792 cm^3 (B) 1781 cm^3 (C) 2771 cm^3 (D) 7125 cm^3 — 1

9 What is the surface area of the closed cylinder in figure 2, to the nearest whole?

(A) 126 cm^2 (B) 252 cm^2 (C) 946 cm^2 (D) 1100 cm^2 — 1

10 Which of these is the area of the trapezium at the front of the prism in figure 3?

(A) 96 cm^2 (B) 108 cm^2 (C) 192 cm^2 (D) 1440 cm^2 — 1

11 What is the volume of the trapezoidal prism in figure 3?

(A) 420 cm^3 (B) 480 cm^3 (C) 960 cm^3 (D) 7200 cm^3 — 1

12 Which of these is the surface area of the trapezoidal prism in figure 3?

(A) 230 cm^2 (B) 330 cm^2 (C) 432 cm^2 (D) 476 cm^2 — 1

13 Which of these is closest to the volume of the cylinder in figure 4?

(A) 140 cm^3 (B) 440 cm^3 (C) 1539 cm^3 (D) 6158 cm^3 — 1

Area, Surface Area and Volume

LEVEL 2 TEST **PART B**

Write answers and working in the space provided.

Marks

14 Find the area of the shaded region in each of these shapes, correct to two decimal places:

a

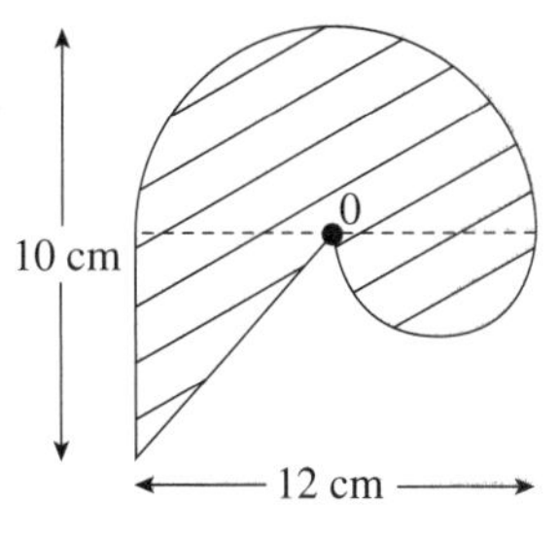

b

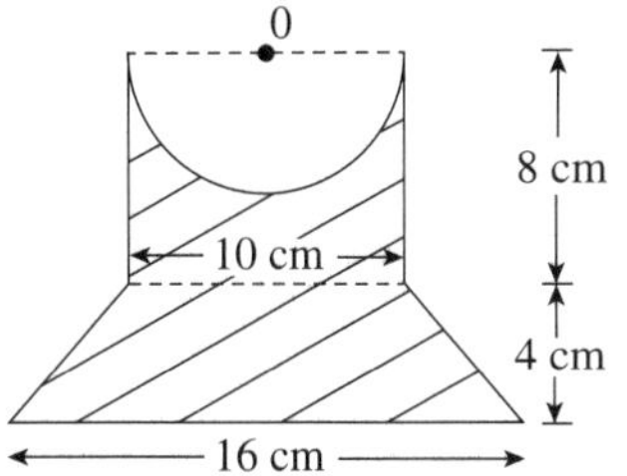

c

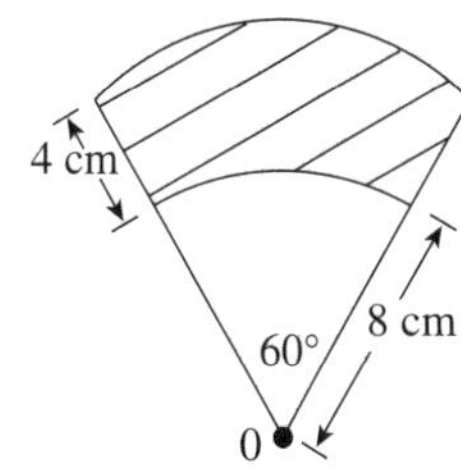

9

The diagrams show three solids which are used to answer questions 15 and 16.

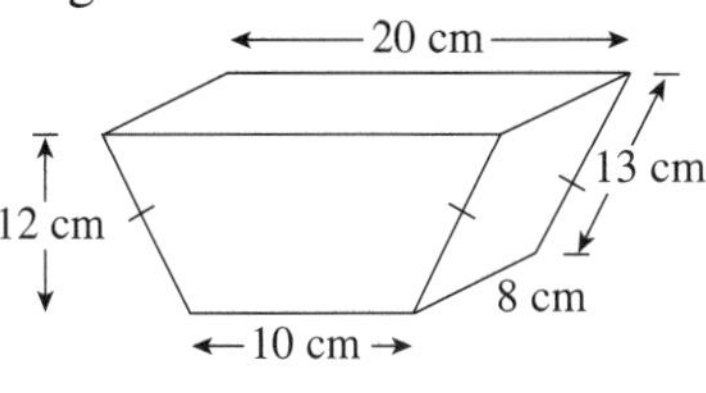

Figure 1

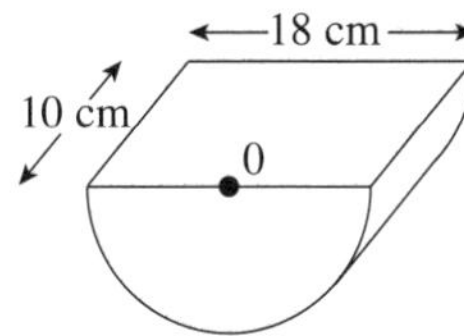

Figure 2

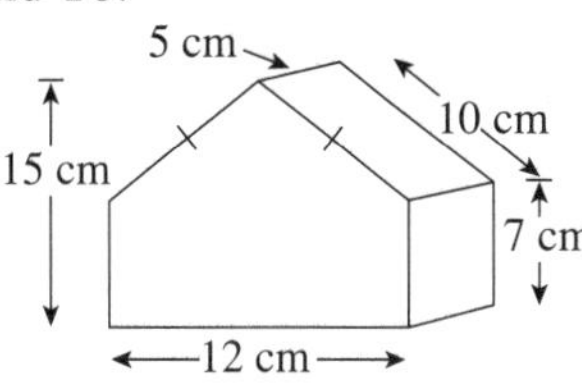

Figure 3

15 Find the surface area of:

a figure 1

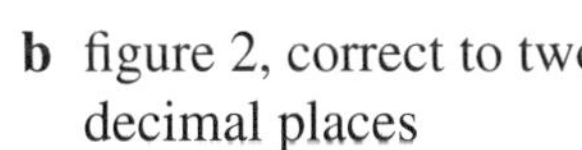

b figure 2, correct to two decimal places

c figure 3

9

16 Find the volume of:

a figure 1

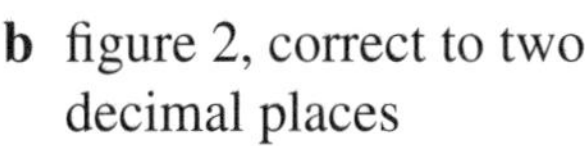

b figure 2, correct to two decimal places

c figure 3

9

Total marks

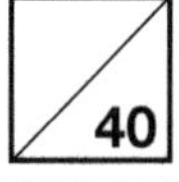

Area, Surface Area and Volume

LEVEL 3 TEST — PART A

Time allowed: 40 minutes

Marks

1 Which of these is closest to the radius of a cylinder with volume 800 cm^3 and height 11 cm?

(A) 4.81 cm (B) 6.38 cm (C) 15.11 cm (D) 23.15 cm — 1

2 What is the volume of a cube with surface area of 61.44 cm^2?

(A) 22.296 cm^3 (B) 32.768 cm^3 (C) 93.423 cm^3 (D) 481.589 cm^3 — 1

A cylinder has a height of 12 cm and diameter of 10 cm. This information is used to answer questions 3 and 4.

3 Which of these is closest to the surface area if the cylinder is **closed**?

(A) 346 cm^2 (B) 534 cm^2 (C) 1382 cm^2 (D) 1659 cm^2 — 1

4 Which of these is closest to the surface area if the cylinder is **open** at both ends?

(A) 377 cm^2 (B) 754 cm^2 (C) 942 cm^2 (D) 1485 cm^2 — 1

5 A rectangular prism with a volume of 576 cm^3 has a length of 12 cm and width of 8 cm. Which of these is the surface area of the prism?

(A) 216 cm^2 (B) 432 cm^2 (C) 448 cm^2 (D) 628 cm^2 — 1

6 A closed cylinder has a diameter of 8 cm and a height of 6 cm. Which of these is the surface area of the cylinder?

(A) 24π cm^2 (B) 48π cm^2 (C) 72π cm^2 (D) 80π cm^2 — 1

7 Which of these is the area of a semicircle with a perimeter of $(6\pi + 12)$ cm?

(A) 6π cm^2 (B) 12π cm^2 (C) 18π cm^2 (D) 24π cm^2 — 1

8 A rectangular pond has dimensions 8 m by 4 m and is surrounded by a 1.5 m wide path. If the path is 10 cm thick, which of these is the volume of the path?

(A) 2.025 m^3 (B) 2.5 m^3 (C) 4.5 m^3 (D) 5.5 m^3 — 1

9 What is the maximum area of a square that can fit inside a circle with area 16π cm^2?

(A) 24 cm^2 (B) 28 cm^2 (C) 32 cm^2 (D) 36 cm^2 — 1

10 A water trough in the shape of a half-cylinder has a capacity of 280 L. If the trough is 2 m long, what is the diameter, to the nearest centimetre?

(A) 30 cm (B) 48 cm (C) 54 cm (D) 60 cm — 1

11 A rectangle with an area of 180 cm^2 is bent to form a cylinder 15 cm high. Which of these is closest to the area of the circular cross-section of the cylinder?

(A) 11 cm^2 (B) 23 cm^2 (C) 113 cm^2 (D) 144 cm^2 — 1

12 The area of an isosceles right triangle is 32 cm^2. Which of these is closest to the perimeter?

(A) 11.3 cm (B) 19.3 cm (C) 21.7 cm (D) 27.3 cm — 1

13 A rectangular garden bed measuring x m by $(x - 2)$ m is surrounded by a 2-m wide section of lawn.

x

$x - 2$

2

Which of these is an expression for the area of the lawn?

(A) $4x$ m^2 (B) $(8x + 8)$ m^2 (C) $(x^2 + 4x + 8)$ m^2 (D) $(x^2 + 8x + 8)$ m^2 — 1

Area, Surface Area and Volume

LEVEL 3 TEST — PART B

Write answers and working in the space provided.

Marks

The diagrams show three solids which are used to answer questions 14 and 15.

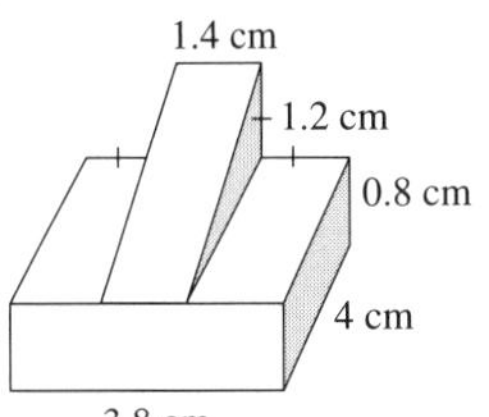

Figure 1

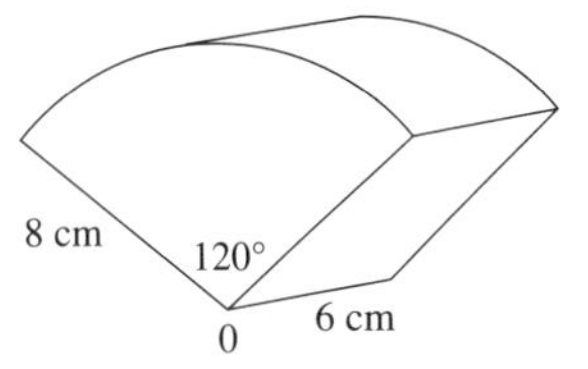

Figure 2

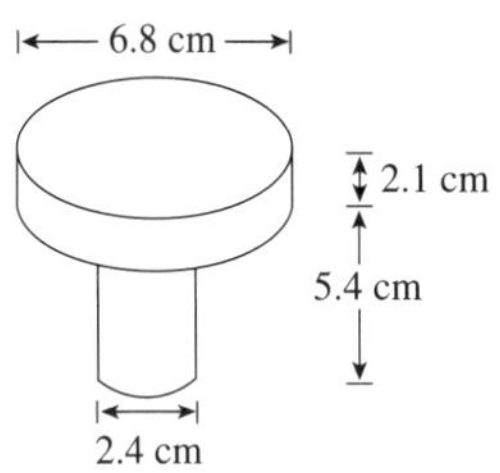

Figure 3

14 Find the volume of:

a figure 1

b figure 2, correct to two decimal places

c figure 3, correct to two decimal places

9

15 Find the surface area of:

a figure 1, correct to two decimal places

b figure 2, correct to two decimal places

c figure 3, correct to two decimal places

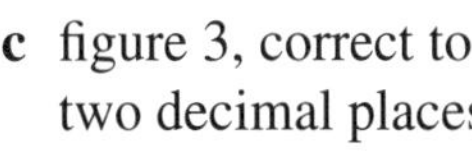

9

16 Find the surface area of each solid, correct to two decimal places if necessary:

a

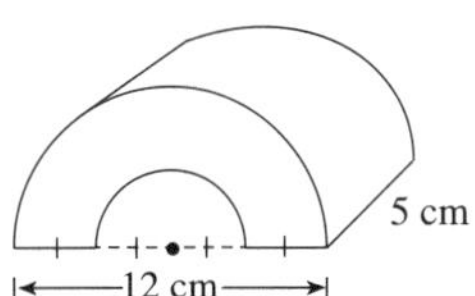

b

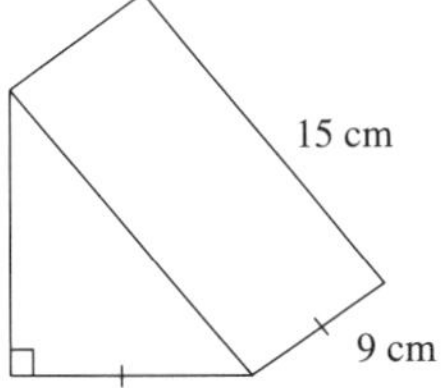

c

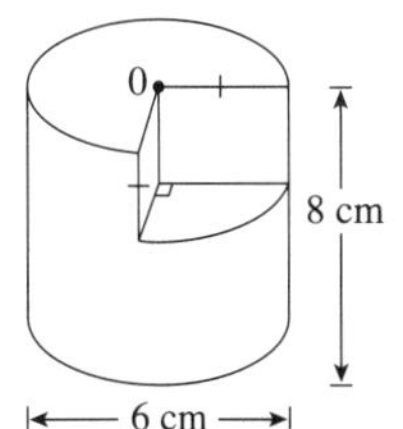

9

Total marks

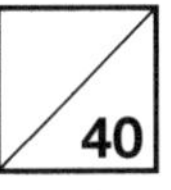

Trigonometry

PRETEST LEVEL 1

Time allowed: 10 minutes **Total marks: 10**

Marks

1 If $\frac{x}{4} = 0.2361$, which of these is the expression for x?

(A) $x = 0.2361 - 4$ (B) $x = 4 \times 0.2361$ (C) $x = \frac{4}{0.2361}$ (D) $x = \frac{0.2361}{4}$

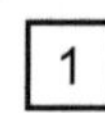

2 Which of these is equal to 16.657 8193 correct to two decimal places?

(A) 16 (B) 16.65 (C) 16.66 (D) 17

The diagrams are used to answer questions 3 to 6.

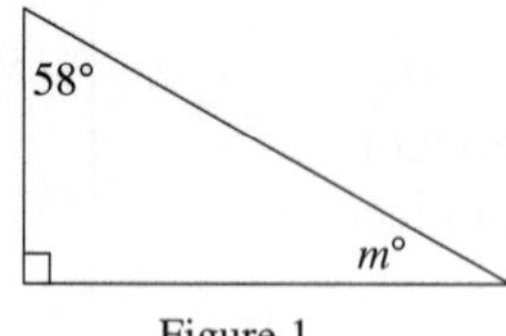

Figure 1

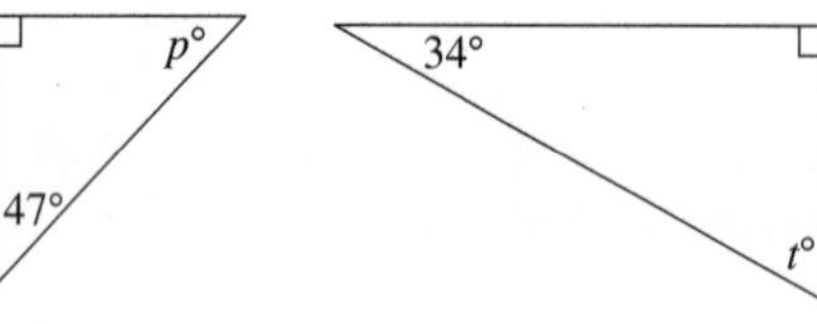

Figure 2 Figure 3

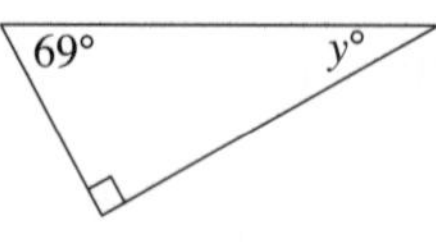

Figure 4

3 In figure 1, what is the value of m?

(A) 32 (B) 42 (C) 58 (D) 122

1

4 Using figure 2, which of these is the value of p?

(A) 47 (B) 43 (C) 53 (D) 133

1

5 Which of these is the value of t in figure 3?

(A) 34 (B) 56 (C) 66 (D) 68

1

6 In figure 4, what is the value of y?

(A) 21 (B) 31 (C) 69 (D) 131

1

7 Which of these is the complement of 60°?

(A) 30° (B) 40° (C) 60° (D) 120°

1

8 If $\frac{y}{6} = 0.906\ 12$, which of these is the value of y correct to two decimal places?

(A) 0.15 (B) 5.43 (C) 5.44 (D) 6.91

9 Which of these is the name given to the side opposite the right angle in a triangle?

(A) complement (B) scalene (C) hypotenuse (D) pythagoras

10

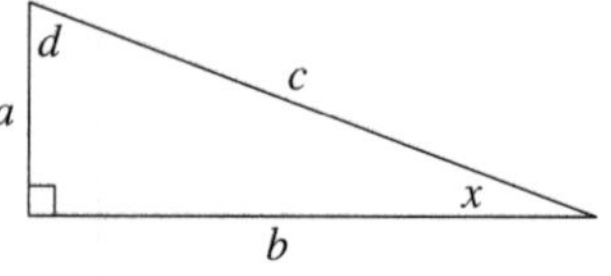

Which of these is the opposite **side** to the angle marked x?

(A) a (B) b (C) c (D) d

Total marks

Trigonometry

PRETEST — LEVELS 2 AND 3

Time allowed: 10 minutes — **Total marks: 10**

Marks

The diagrams are used to answer questions 1 to 4.

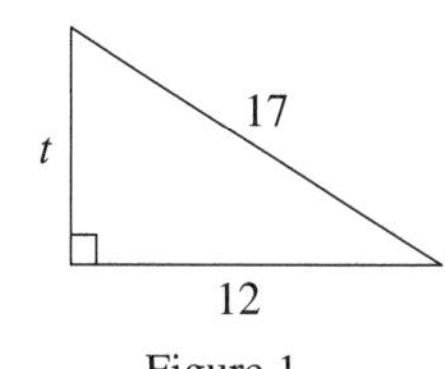

Figure 1

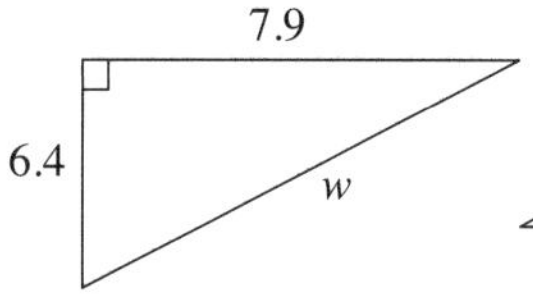

Figure 2

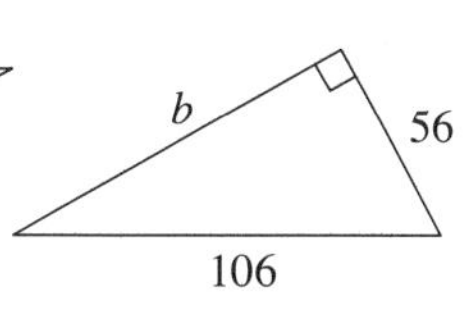

Figure 3

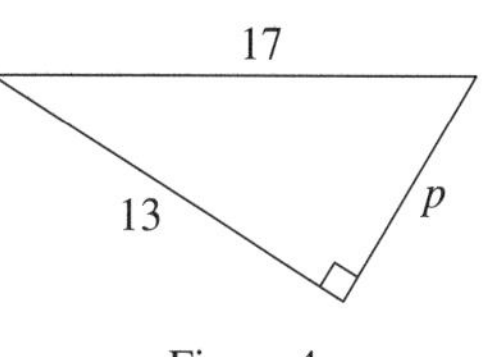

Figure 4

1 In figure 1, what is the value of t, correct to two decimal places?

(A) 5.00 (B) 12.04 (C) 12.05 (D) 12.06

1

2 In figure 2, what is the value of w, correct to three significant figures?

(A) 10.1 (B) 10.16 (C) 10.17 (D) 10.2

1

3 In figure 3, which of these is the exact value of b?

(A) $\sqrt{50}$ (B) $\sqrt{162}$ (C) 89 (D) 90 — 1

4 In figure 4, which of these is the closest to the exact value of p?

(A) 10.9 (B) 10.95 (C) 10.96 (D) 11 — 1

5 Which of these is closest to the value of x if $\frac{4}{x} = 0.48$?

(A) 0.1 (B) 1.9 (C) 8.3 (D) 8.4 — 1

6 Which of these is the value of a if $\frac{a}{\sqrt{5}} = 6$, correct to three decimal places?

(A) 5.477 (B) 8.236 (C) 11.011 (D) 13.416 — 1

The diagrams are used to answer questions 7 to 10.

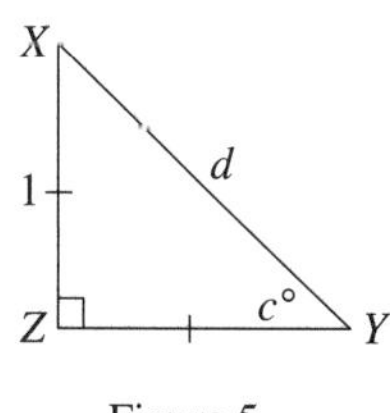

Figure 5

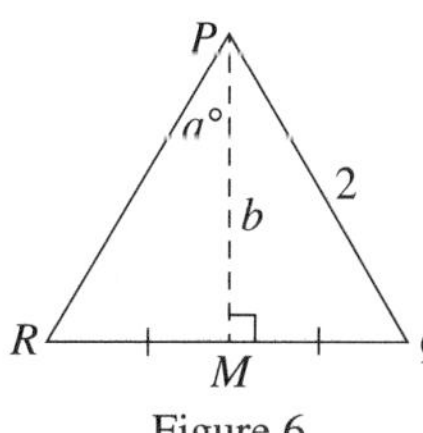

Figure 6

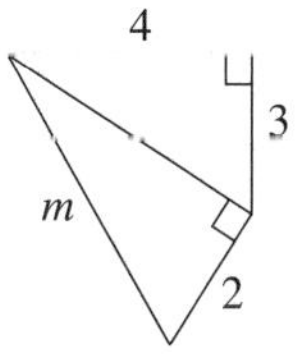

Figure 7

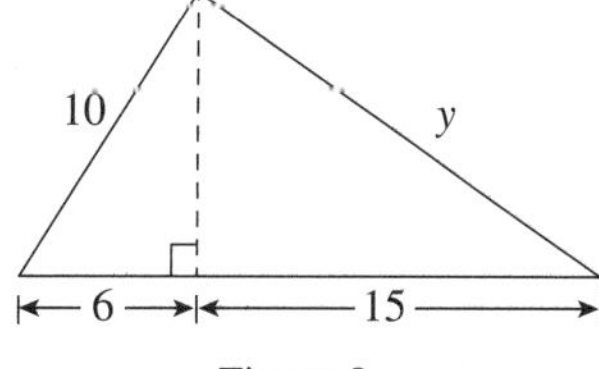

Figure 8

7 In figure 5, triangle XYZ is isosceles where $XZ = YZ$. Which of these is true?

(A) $c = 30, d = \sqrt{2}$ (B) $c = 30, d = 2$ (C) $c = 45, d = \sqrt{2}$ (D) $c = 45, d = 2$ — 1

8 In figure 6, triangle PQR is equilateral and $RM = QM$. Which of these is true?

(A) $a = 30, b = \sqrt{3}$ (B) $a = 30, b = \sqrt{5}$ (C) $a = 45, b = \sqrt{3}$ (D) $a = 45, b = \sqrt{5}$

1

9 In figure 7, what is the length of m?

(A) 5 units (B) $\sqrt{27}$ units (C) $\sqrt{28}$ units (D) $\sqrt{29}$ units — 1

10 In figure 8, what is the value of y?

(A) 17 (B) 18 (C) 21 (D) 23

1

Total marks /10

Trigonometry

LEVEL 1 TEST — PART A

Time allowed: 40 minutes

Marks

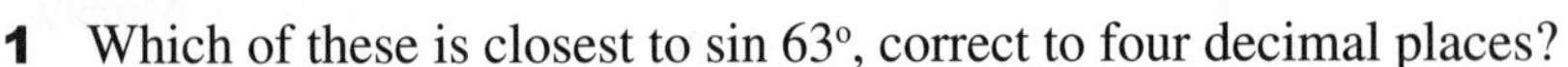

1 Which of these is closest to sin 63°, correct to four decimal places?

(A) 0.1673 (B) 0.1674 (C) 0.8910 (D) 0. 8911 — 1

2 What is the value of 9 × tan 20°, correct to two decimal places?

(A) 3.27 (B) 3.28 (C) 20.13 (D) 20.14 — 1

3 What is the value of 6 ÷ tan 42°, correct to the nearest whole?

(A) 2 (B) 3 (C) 6 (D) 7 — 1

The diagrams are used to answer questions 4 to 13.

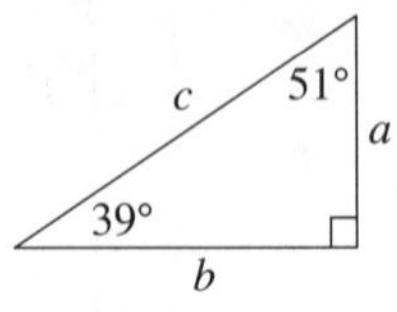

Figure 1

m
36°
n
p
54°

Figure 2

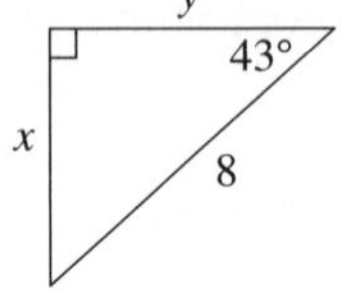

Figure 3

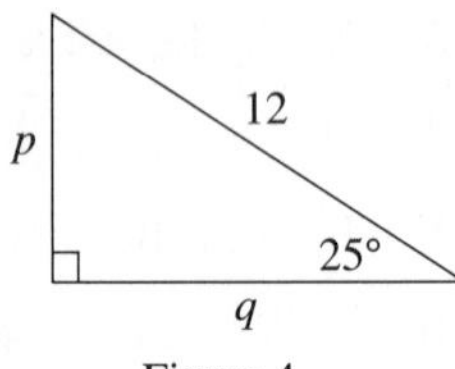

Figure 4

4 In figure 1, which of these is the side opposite to the angle measuring 39°?

(A) a (B) b (C) c (D) none of these — 1

5 In figure 1, which of these is the side adjacent to the angle measuring 39°?

(A) a (B) b (C) c (D) none of these — 1

6 In figure 1, which of these is the hypotenuse?

(A) a (B) b (C) c (D) none of these — 1

7 In figure 2, which of these is equal to sin 54°?

(A) $\frac{m}{n}$ (B) $\frac{n}{p}$ (C) $\frac{p}{m}$ (D) $\frac{m}{p}$ — 1

8 In figure 2, which of these is equal to cos 54°?

(A) $\frac{n}{p}$ (B) $\frac{n}{m}$ (C) $\frac{m}{p}$ (D) $\frac{p}{m}$ — 1

9 In figure 2, which of these is equal to tan 36°?

(A) $\frac{n}{m}$ (B) $\frac{n}{p}$ (C) $\frac{m}{n}$ (D) $\frac{m}{p}$ — 1

10 In figure 3, which of these is equal to the expression for the value of x?

(A) $x = 8 \times \sin 43°$ (B) $x = 8 \times \cos 43°$ (C) $x = 8 \times \tan 43°$ (D) $x = \frac{8}{\cos 43°}$ — 1

11 In figure 3, which of these is equal to the expression for the value of y?

(A) $y = \frac{8}{\tan 43°}$ (B) $y = \frac{\tan 43°}{8}$ (C) $y = 8 \times \cos 43°$ (D) $y = 8 \times \tan 43°$ — 1

12 In figure 4, which of these is equal to the expression for the value of p?

(A) $p = 25 \times \sin 12°$ (B) $p = 12 \times \sin 25°$ (C) $p = 12 \times \tan 25°$ (D) $p = 25 \times \tan 12°$ — 1

13 In figure 4, which of these is equal to the expression for the value of q?

(A) $q = \frac{12}{\cos 25°}$ (B) $q = 12 \times \tan 25°$ (C) $q = 12 \times \sin 25°$ (D) $q = 12 \times \cos 25°$ — 1

Trigonometry

LEVEL 1 TEST — PART B

Write answers and working in the space provided.

14 Find the value of the pronumeral in each of the triangles, correct to two decimal places:

a

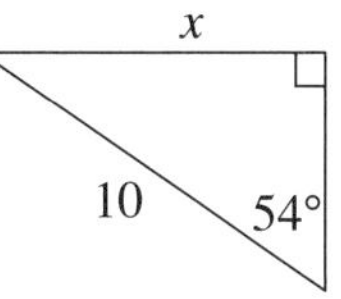

b

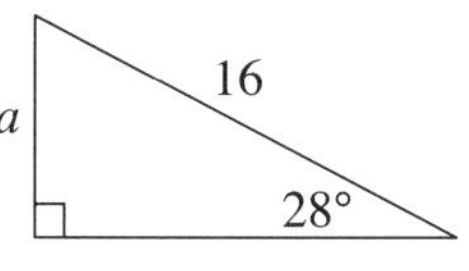

c

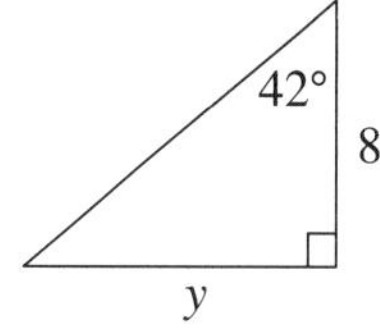

Marks: 6

d

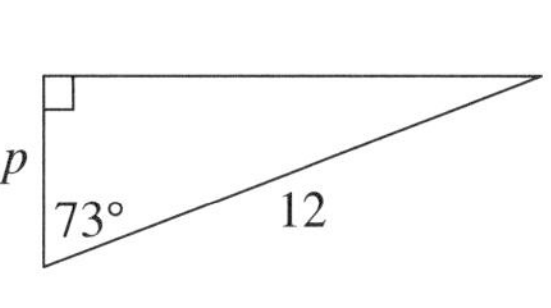

e

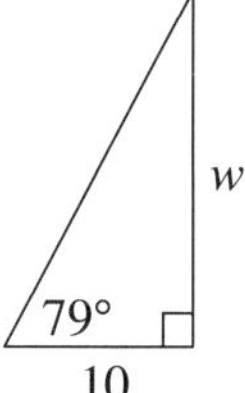

f

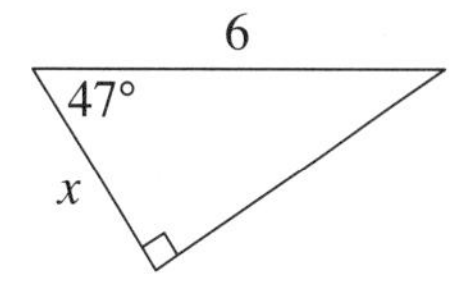

Marks: 6

15 Find the value of θ, to the nearest whole, given that θ is acute:

a $\sin \theta = 0.4531$ **b** $\cos \theta = 0.7814$ **c** $\tan \theta = 2.4518$

Marks: 3

16 Find the size of the angle θ, to the nearest degree:

a

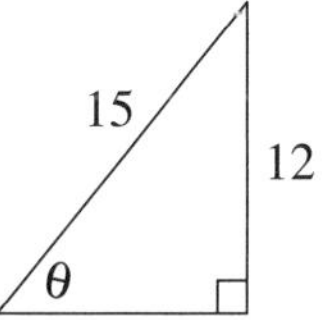

b

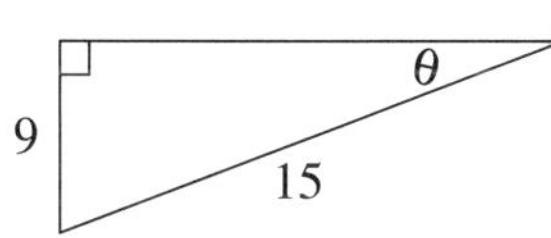

c

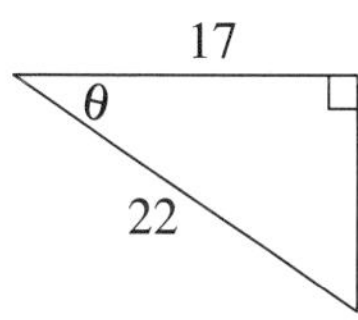

Marks: 6

d

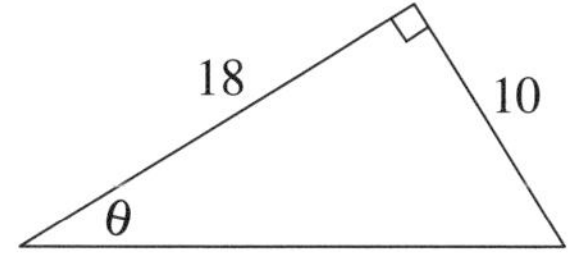

e

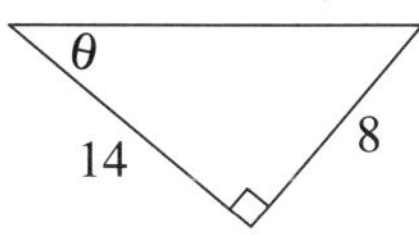

f

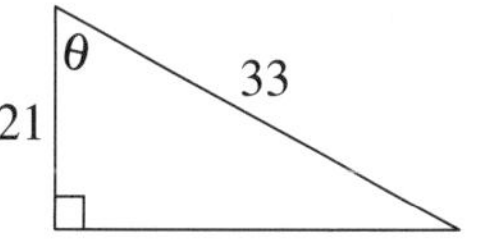

Marks: 6

Total marks

Trigonometry

LEVEL 2 TEST **PART A**

Time allowed: 40 minutes

Marks

1 Which of these is the value of $\frac{\sin 45°}{\cos 45°}$?

Ⓐ 0.01 Ⓑ 0.1 Ⓒ 0.5 Ⓓ 1 — 1

2 When θ is acute, which of these is closest to the value of θ when $\sin \theta = 0.412$?

Ⓐ 24° Ⓑ 25° Ⓒ 65° Ⓓ 66° — 1

3 Which of these is tan 18°15′, correct to two decimal places?

Ⓐ 0.26 Ⓑ 0.27 Ⓒ 0.32 Ⓓ 0.33 — 1

4 What is the value of x when $\sin 40° = \cos x°$?

Ⓐ $x = 20$ Ⓑ $x = 40$ Ⓒ $x = 50$ Ⓓ $x = 140$ — 1

5 Which of these is closest to 38.495°?

Ⓐ 38°29′ Ⓑ 38°30′ Ⓒ 38°49′ Ⓓ 38°50′ — 1

6 Which of these is closest to the value of θ if $\tan \theta = \frac{2}{\sqrt{5}}$, given θ is acute?

Ⓐ 28°22′ Ⓑ 28°37′ Ⓒ 41°48′ Ⓓ 41°49′ — 1

The diagrams are used to answer questions 7 to 13.

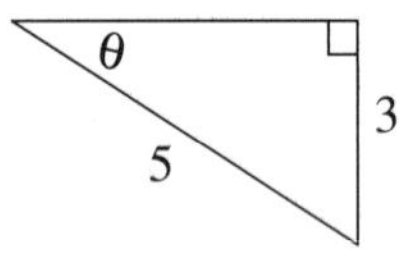

Figure 1

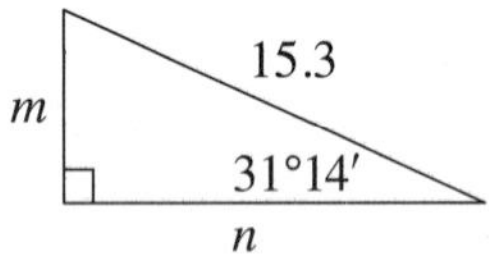

m
15.3
31°14′
n

Figure 2

y
43°
5.1
x

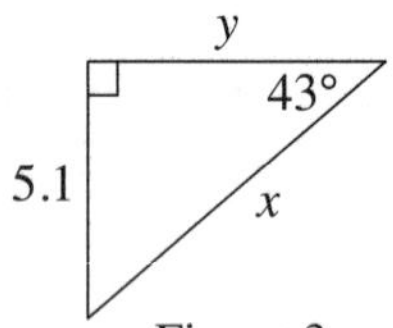

Figure 3

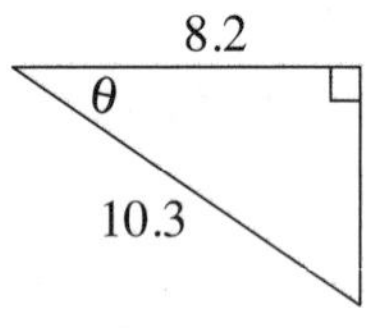

Figure 4

7 Using figure 1 and Pythagoras' theorem, which of these is the value of $\cos \theta$?

Ⓐ $\frac{3}{5}$ Ⓑ $\frac{3}{4}$ Ⓒ $\frac{4}{5}$ Ⓓ $\frac{4}{3}$ — 1

8 Using figure 1, what is the size of θ, to the nearest minute?

Ⓐ 36°8′ Ⓑ 36°52′ Ⓒ 53°15′ Ⓓ 53°8′ — 1

9 Using figure 2, what is the value of m, correct to two decimal places?

Ⓐ 7.93 Ⓑ 8.16 Ⓒ 8.93 Ⓓ 13.08 — 1

10 Using figure 2, what is the value of n, correct to three significant figures?

Ⓐ 11.4 Ⓑ 12.9 Ⓒ 13.1 Ⓓ 13.2 — 1

11 Using figure 3, which of these is the expression for the value of x?

Ⓐ $x = \frac{5.1}{\sin 43°}$ Ⓑ $x = \frac{5.1}{\cos 43°}$ Ⓒ $x = \frac{\sin 43°}{5.1}$ Ⓓ $x = \frac{\cos 43°}{5.1}$ — 1

12 Using figure 3, which of these is the value of y, correct to two decimal places?

Ⓐ 4.74 Ⓑ 4.75 Ⓒ 5.46 Ⓓ 5.47 — 1

13 Using figure 4, which of these is the value of θ, correct to the nearest minute?

Ⓐ 37°14′ Ⓑ 37°15′ Ⓒ 37°23′ Ⓓ 37°24′ — 1

Trigonometry

LEVEL 2 TEST

PART B

Write answers and working in the space provided.

Marks

14 a

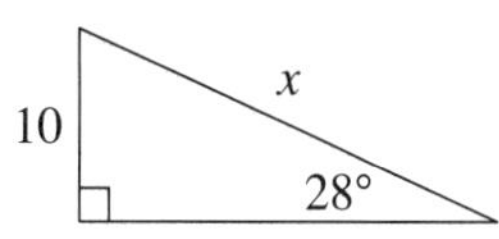

Find x, to two decimal places.

b

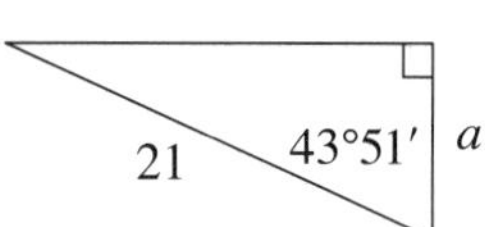

Find a, correct to two decimal places.

c

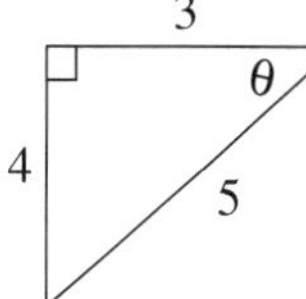

Find θ to the nearest minute.

6

d

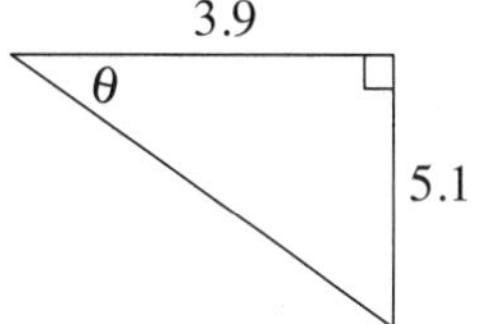

Find θ to the nearest minute.

e

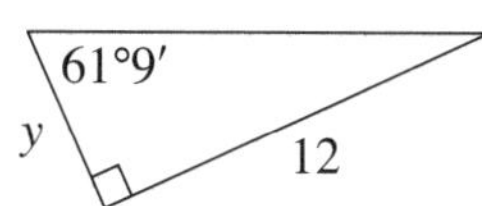

Find y, correct to two decimal places.

f

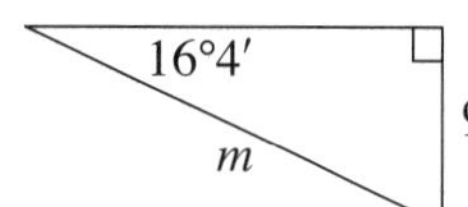

Find m, correct to two decimal places.

6

15 a When the altitude of the sun is 32°16′ a vertical stick casts a shadow 3.1 m long. How high is the stick, correct to two decimal places?

b A ladder leans against a wall and makes an angle of 64° with the floor. If the ladder is 2.4 m long, how far does it reach up the wall, to the nearest centimetre?

c The diagonal of a rectangle is 14.6 cm in length and makes an angle of 32° with the longer side. What is the length of the longer side of the rectangle, correct to two decimal places?

9

16 a If $\sin\theta = \frac{4}{5}$ and θ is acute, what is the value of $\cos\theta$?

b If $\cos\theta = \frac{12}{13}$ and θ is acute, what is the value of $\tan\theta$?

c If $\tan\theta = \frac{\sqrt{2}}{3}$ and θ is acute, what is the value of $\cos\theta$?

6

Total marks

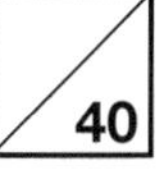
40

Trigonometry

LEVEL 3 TEST — PART A

Time allowed: 40 minutes

Marks

The diagrams show three figures and are used to answer questions 1 to 6.

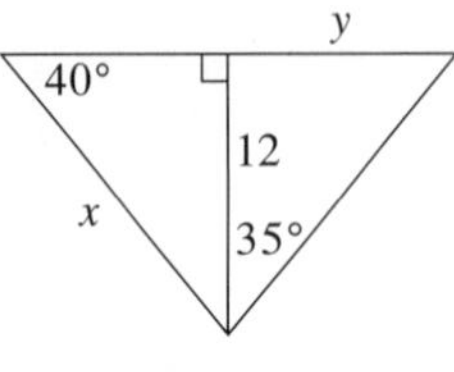

Figure 1

Figure 2

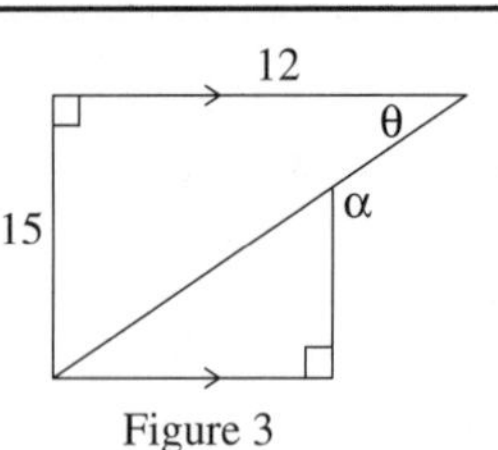

Figure 3

1 Using figure 1, which of these is the value of x, correct to two decimal places?

Ⓐ 7.71 Ⓑ 9.19 Ⓒ 15.66 Ⓓ 18.67 1

2 Using figure 1, which of these is the value of y, correct to two decimal places?

Ⓐ 6.88 Ⓑ 8.40 Ⓒ 9.83 Ⓓ 17.14 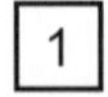

3 Using figure 2, which of these is the value of m, correct to three significant places?

Ⓐ 18.8 Ⓑ 19.37 Ⓒ 20.9 Ⓓ 21.5 1

4 Using figure 2, which of these is the value of n, correct to three significant places?

Ⓐ 15.3 Ⓑ 15.9 Ⓒ 17.3 Ⓓ 17.9 1

5 Using figure 3, which of these is the size of angle θ, to the nearest degree?

Ⓐ 37° Ⓑ 39° Ⓒ 48° Ⓓ 51° 1

6 Using figure 3, which of these is the size of angle α, to the nearest degree?

Ⓐ 127° Ⓑ 129° Ⓒ 138° Ⓓ 141° 1

The diagrams show three figures and are used to answer questions 7 to 13.

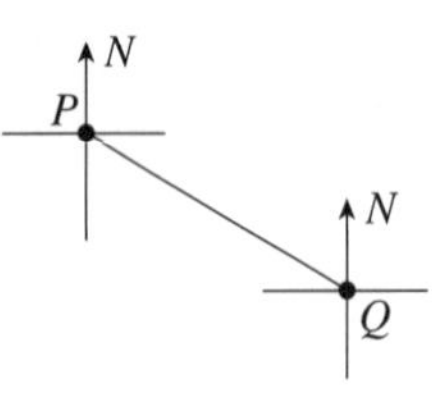

Figure 4

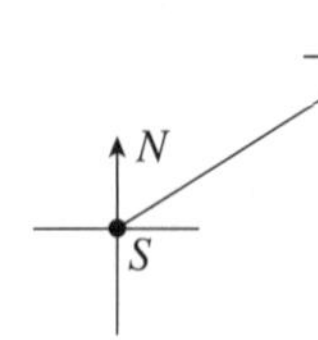

Figure 5

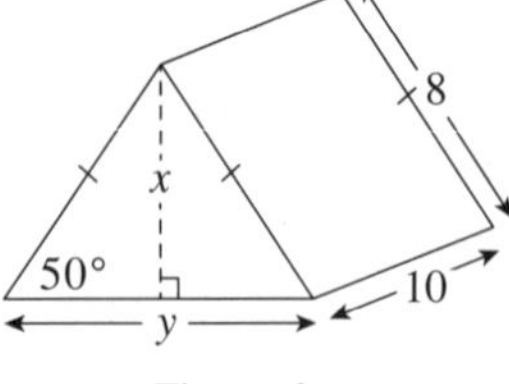

Figure 6

7 If the bearing of Q from P is 130°, which of these is the bearing of P from Q?

Ⓐ 50° Ⓑ 130° Ⓒ 310° Ⓓ 320° 1

8 If the bearing of P from Q is x°, which of these is the bearing of Q from P?

Ⓐ $(180 - x)°$ Ⓑ $(x - 180)°$ Ⓒ $(90 + x)°$ Ⓓ $(270 - x)°$ 1

9 If the bearing of S from R is 220°, which of these is the bearing of R from S?

Ⓐ 040° Ⓑ 050° Ⓒ 060° Ⓓ 080° 1

10 If the bearing of S from R is y°, which of these is the bearing of R from S?

Ⓐ $(y - 180)°$ Ⓑ $(180 - y)°$ Ⓒ $(180 + y)°$ Ⓓ $(270 - y)°$ 1

11 Using figure 6, which of these is the value of x, correct to three significant places?

Ⓐ 5.14 Ⓑ 6.13 Ⓒ 10.4 Ⓓ 12.4 1

12 Using figure 6, which of these is the value of y, correct to three significant places?

Ⓐ 5.14 Ⓑ 6.13 Ⓒ 10.3 Ⓓ 12.3 1

13 Using figure 6, which of these is closest to the volume of the triangular prism?

Ⓐ 158 units^3 Ⓑ 231 units^3 Ⓒ 315 units^3 Ⓓ 463 units^3 1

LEVEL 3 TEST

PART B

Write answers and working in the space provided.

Marks

14 a Find x and y to two decimal places.

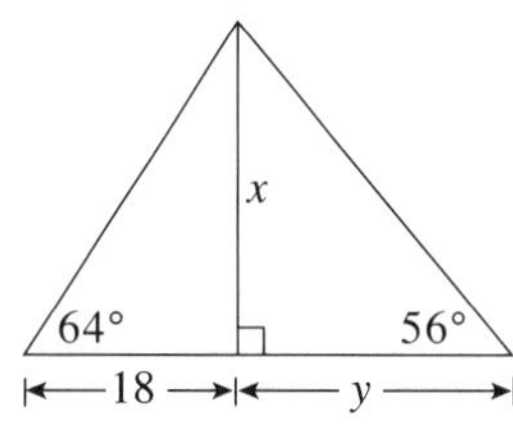

b Find θ to the nearest minute.

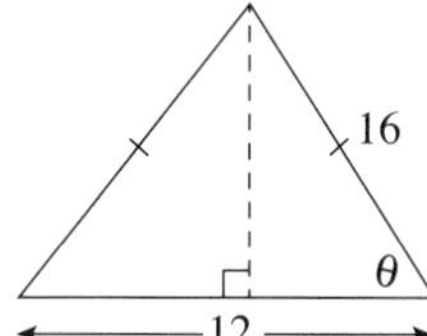

c Find the perimeter to one decimal place.

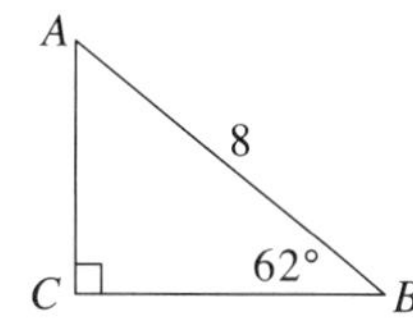

9

15 a Find x to two decimal places.

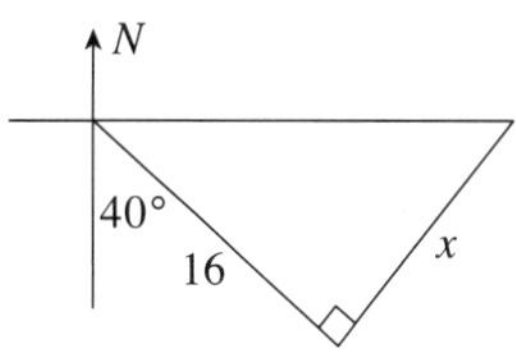

b Find θ to the nearest degree.

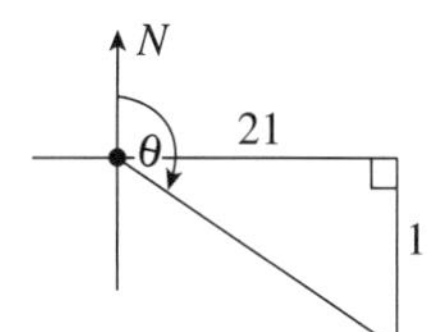

c Find the bearing of C from A.

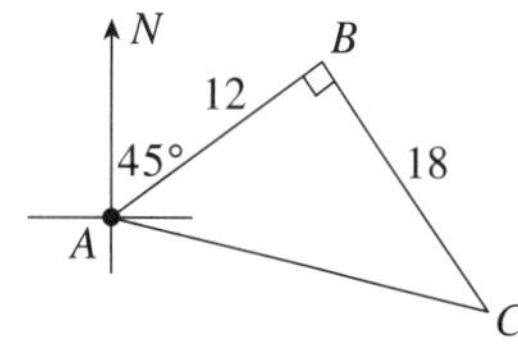

9

16 a Find θ to the nearest degree.

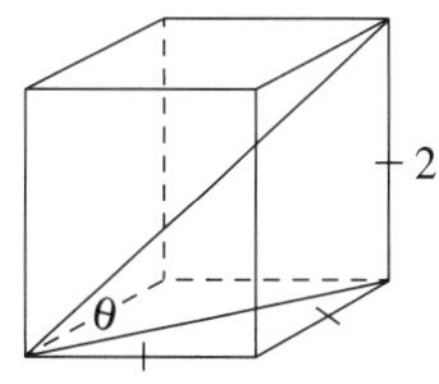

b Find θ to the nearest degree.

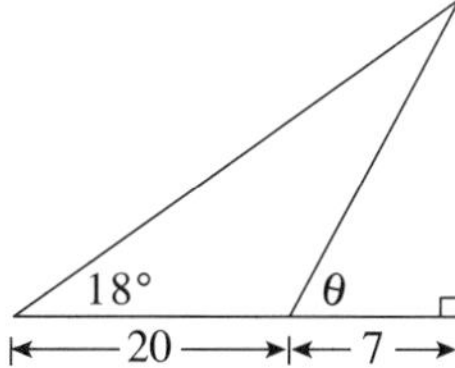

c Find θ to the nearest minute.

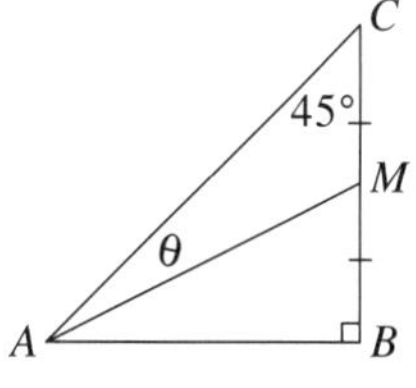

9

Total marks

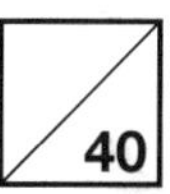

PRETEST

LEVEL 1

Time allowed: 10 minutes **Total marks: 10**

Marks

The diagrams are used to answer questions 1 to 4.

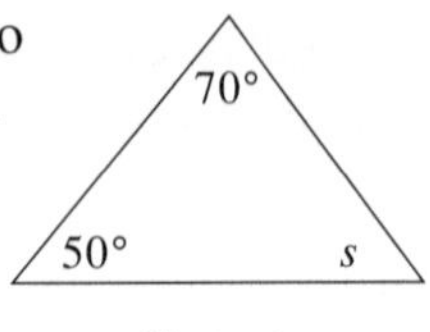

Figure 1

3
w

Figure 2

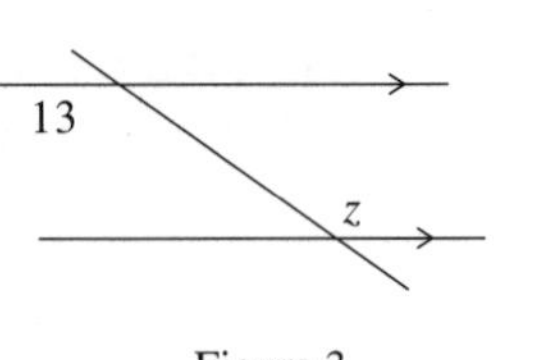

Figure 3

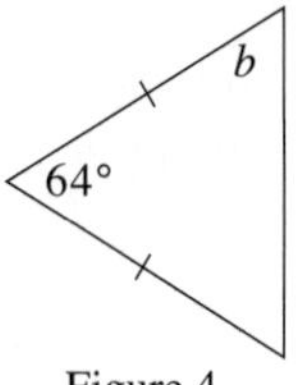

Figure 4

1 In figure 1, which of these is the value of *s*?

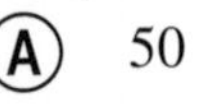

Ⓐ 50   Ⓑ 60   Ⓒ 70   Ⓓ 80   1

2 In figure 2, which of these is the value of *w*?

Ⓐ 30   Ⓑ 50   Ⓒ 60   Ⓓ 70   1

3 In figure 3, which of these is the value of *z*?

Ⓐ 50   Ⓑ 65   Ⓒ 130   Ⓓ 150   1

4 In figure 4, which of these is the value of *b*?

Ⓐ 32   Ⓑ 58   Ⓒ 64   Ⓓ 116   1

5 Which of these is the same as 8:18?

Ⓐ 4:9   Ⓑ 2:3   Ⓒ 3:2   Ⓓ 9:4   1

The diagram shows a rectangle *ABCD* and a right-angled isosceles triangle *DEC*, and is used to answer questions 6 to 8.

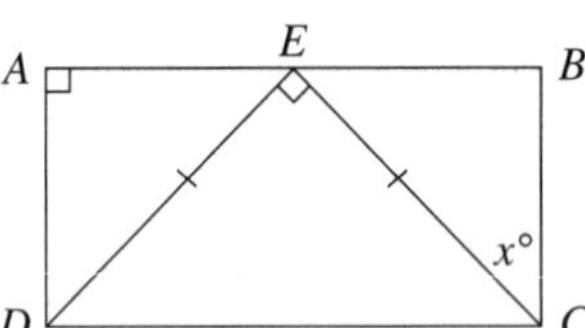

6 Which of these angles is the same size as angle *EDC*?

Ⓐ *EAD*   Ⓑ *EBC*   Ⓒ *BCD*   Ⓓ *ECD*   1

7 Which of these sides is the same length as side *AD*?

Ⓐ *BC*   Ⓑ *ED*   Ⓒ *AB*   Ⓓ *DC*   1

8 Which of these is the angle marked as $x°$?

Ⓐ *DCE*   Ⓑ *ECB*   Ⓒ *BEA*   Ⓓ *BCD*   1

The diagram shows two triangles and is used to answer questions 9 and 10.

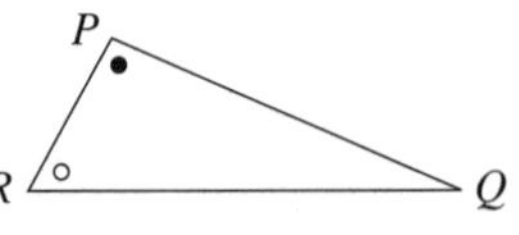

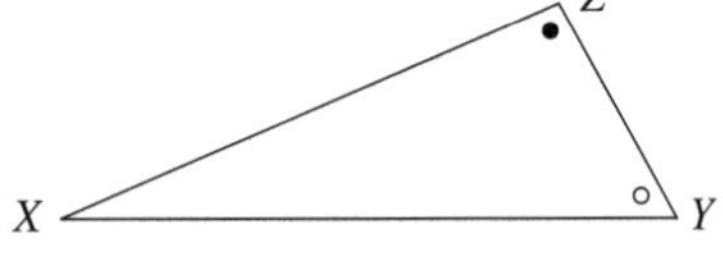

9 Which of these is equal to angle *PRQ*?

Ⓐ *RPQ*   Ⓑ *XZY*   Ⓒ *YXZ*   Ⓓ *ZYX*   1

10 Which of these is equal to angle *YXZ*?

Ⓐ *RQP*   Ⓑ *QPR*   Ⓒ *PRQ*   Ⓓ *XYZ*   1

Total marks

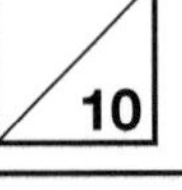

Similarity

PRETEST

LEVELS 2 AND 3

Time allowed: 10 minutes **Total marks: 10**

Marks

1 If $3:8 = 9:a$, which of these is the value of a?

(A) 11 (B) 14 (C) 24 (D) 27 [1]

2 Which of these is the solution to the equation $\frac{x}{5} = 10$?

(A) $x = 2$ (B) $x = 5$ (C) $x = 15$ (D) $x = 50$ [1]

3 If three milkshakes cost \$9.60, what is the cost of four milkshakes?

(A) \$3.20 (B) \$12.80 (C) \$13.60 (D) \$199.20 [1]

4 A rectangle has dimensions 5 cm by 2 cm. If the dimensions are doubled, what is the area of the new rectangle?

(A) 20 cm^2 (B) 40 cm^2 (C) 50 cm^2 (D) 100 cm^2 [1]

5 A plant is now five times the height it was when it was planted in a garden. If the plant is now 85 cm, what was the height when it was planted?

(A) 17 cm (B) 20 cm (C) 80 cm (D) 90 cm [1]

6 Which of these is the solution of $\frac{x}{4} = \frac{3}{5}$?

(A) $x = 2.4$ (B) $x = 2.5$ (C) $x = 2.6$ (D) $x = 2.8$ [1]

In the diagram ΔABC has been enlarged, rotated and translated to produce ΔPQR. The diagram is used to answer questions 7 and 8.

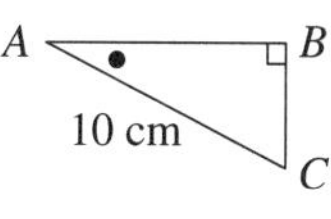

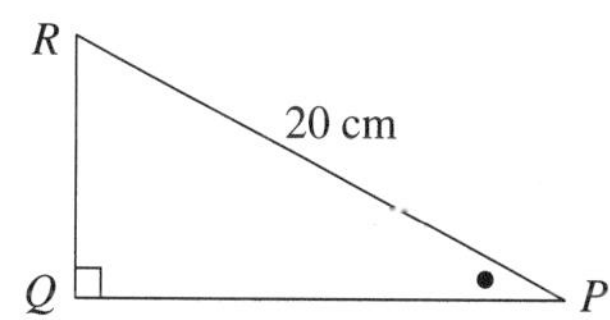

7 Which of these is **not** a correct statement?

(A) The length of PR is twice the length of AC.
(B) The length of AB is twice the length of PQ.
(C) The length of AB is half the length of PQ.
(D) The length of BC is half the length of QR. [1]

8 If the perimeter of ΔABC is p cm, which of these is the perimeter of ΔPQR?

(A) $2p$ cm (B) $(p + 2)$ cm (C) $3p$ cm (D) $6p$ cm [1]

9 Here is a pattern of numbers: $\frac{1}{2}, \frac{3}{8}, \frac{9}{32}$ ... Which of these is the next number in the pattern?

(A) $\frac{3}{4}$ (B) $\frac{12}{40}$ (C) $\frac{27}{64}$ (D) $\frac{27}{128}$ [1]

10 Which of these is the solution of $\frac{4}{0.5} = \frac{24}{m}$?

(A) $m = 3$ (B) $m = 4$ (C) $m = 6$ (D) $m = 8$ [1]

Total marks /10

Similarity

LEVEL 1 TEST

PART A

Time allowed: 40 minutes

Marks

The diagrams show three similar figures and are used to answer questions 1 to 8.

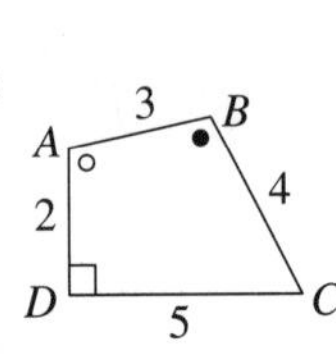

Figure 1

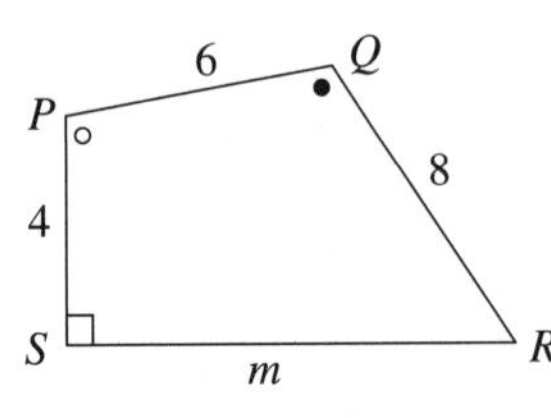

Figure 2

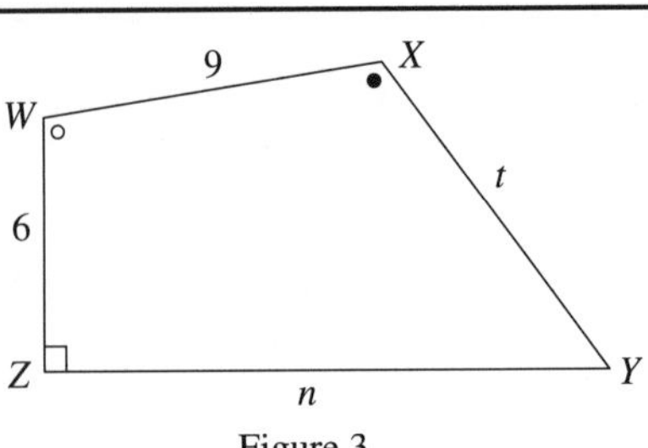

Figure 3

1 If figure 1 is the original and figure 2 is its image, what is the enlargement factor?

Ⓐ 2 Ⓑ 3 Ⓒ 4 Ⓓ 8 | 1

2 If figure 1 is the original and figure 3 is its image, what is the enlargement factor?

Ⓐ 2 Ⓑ 3 Ⓒ 4 Ⓓ 8 | 1

3 Using figure 1 and figure 2, what is the value of m?

Ⓐ 9 Ⓑ 10 Ⓒ 12 Ⓓ 15 | 1

4 Using figure 1 and figure 3, what is the value of n?

Ⓐ 9 Ⓑ 10 Ⓒ 12 Ⓓ 15 | 1

5 Which of these is the value of t?

Ⓐ 9 Ⓑ 10 Ⓒ 12 Ⓓ 15 | 1

6 Which of these is a matching side of PS?

Ⓐ WZ Ⓑ AB Ⓒ ZY Ⓓ PQ | 1

7 Which of these is a matching side of XW?

Ⓐ BC Ⓑ WZ Ⓒ BA Ⓓ XY | 1

8 Which of these is a matching angle of BCD?

Ⓐ XYZ Ⓑ WZY Ⓒ CDA Ⓓ QPS | 1

A map is drawn using a scale of 1 cm = 5 km and is used to answer questions 9 to 11.

9 Two towns are 10 cm apart on the map. Which of these is the real distance between them?

Ⓐ 2 km Ⓑ 5 km Ⓒ 20 km Ⓓ 50 km | 1

10 The distance between two intersections is 120 km. What is the distance between the two intersections on the map?

Ⓐ 12 cm Ⓑ 24 cm Ⓒ 40 cm Ⓓ 115 cm | 1

11 Which of these is the scale of the map written as a ratio?

Ⓐ 1 : 5 Ⓑ 1 : 500 Ⓒ 1 : 100 000 Ⓓ 1 : 500 000 | 1

12 A photograph is 15 cm long and 10 cm wide. The photograph is to be enlarged so that the new photo will be 30 cm wide. What is the new length?

Ⓐ 5 cm Ⓑ 20 cm Ⓒ 45 cm Ⓓ 50 cm | 1

13 Each side of an equilateral triangle is 6 cm. The triangle is to be enlarged by a factor of 2. Which of these will be the perimeter of the new triangle?

Ⓐ 12 cm Ⓑ 18 cm Ⓒ 24 cm Ⓓ 36 cm | 1

Similarity

LEVEL 1 TEST — PART B

Write answers and working in the space provided.

Marks

14 The diagrams show pairs of similar figures:

a

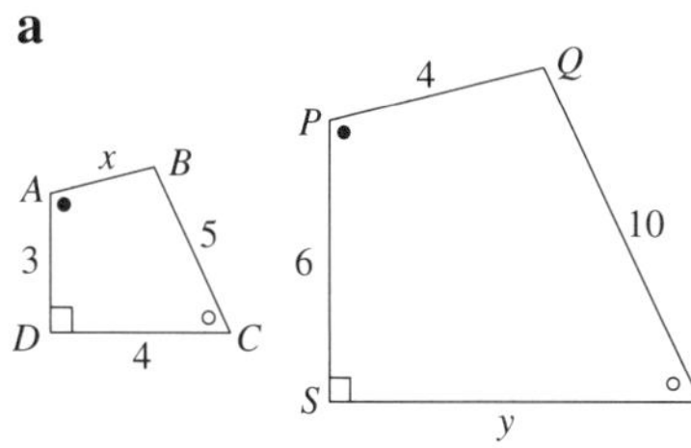

a i What is the enlargement factor? ______ **ii** What side matches DC? ______

iii What angle matches PQR? ______ **iv** Complete $\frac{PS}{AD} = \frac{\quad}{AB}$

v What is the value of x? ______ **vi** What is the value of y? ______

6

b

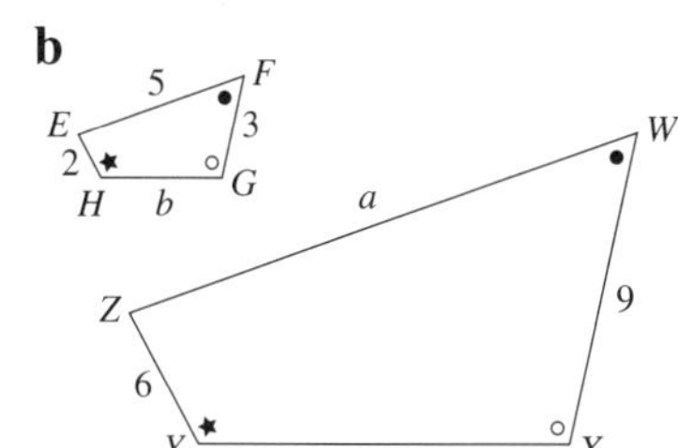

b i What is the enlargement factor? ______ **ii** What side matches WZ? ______

iii What angle matches HEF? ______ **iv** Complete  $\frac{XW}{GF} = \frac{\quad}{EH}$

v What is the value of a? ______ **vi** What is the value of b? ______

6

15 Decide whether the following pairs are similar:

a

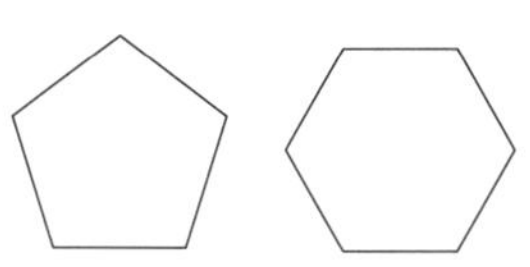

b

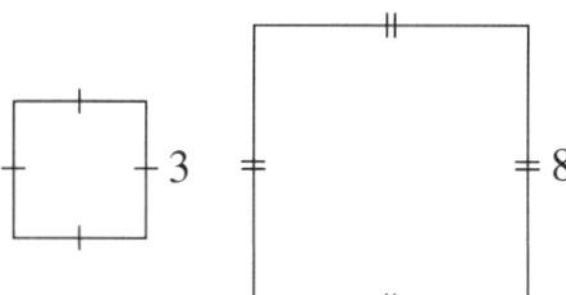

c

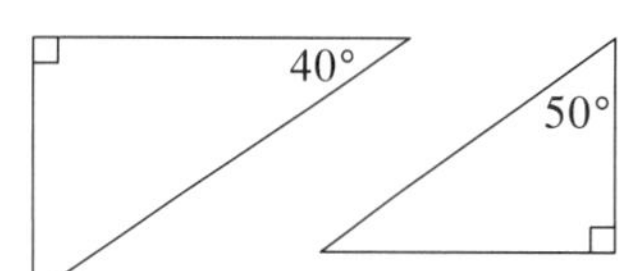

______ ______ ______

3

16 Find the value of the pronumerals in the following pairs of similar shapes:

a

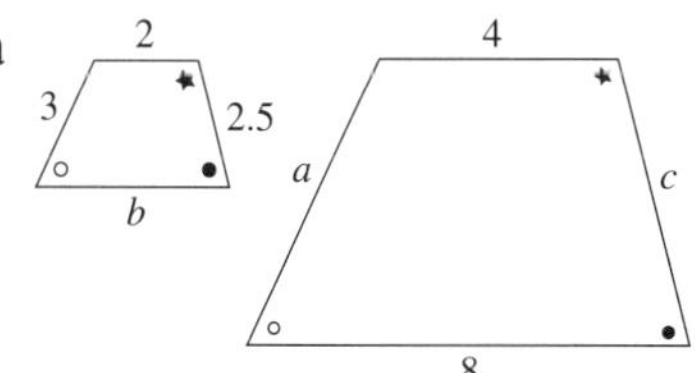

a i a ______

ii b ______

iii c ______

3

b

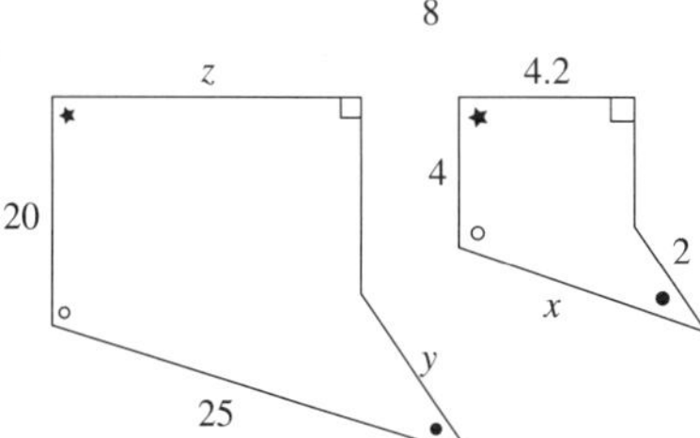

b i x ______

ii y ______

iii z ______

3

17 At a particular time a 1-m high stick casts a shadow of 1.5 m. At that same time, what is the:

a length of shadow of a tree 8 m high?

b length of shadow of a flagpole 6.4 m tall?

c height of a building that casts a shadow 6 m?

______ ______ ______

6

Total marks /40

Similarity

LEVEL 2 TEST PART A

Time allowed: 40 minutes

Marks

A scale drawing has a scale of 1:2000 and is used to answer questions 1 to 3.

1 Which of these is the scale used on the drawing?

(A) 1 cm = 2 m (B) 1 cm = 20 m (C) 1 cm = 200 m (D) 1 cm = 2000 m 1

2 What length on the drawing represents a distance of 8 m?

(A) 4 mm (B) 16 mm (C) 4 cm (D) 16 cm 1

3 What distance is represented by a distance on the drawing of 6.8 cm?

(A) 3.4 m (B) 13.6 m (C) 34 m (D) 136 m 1

The diagrams show pairs of similar triangles and are used to answer questions 4 to 7.

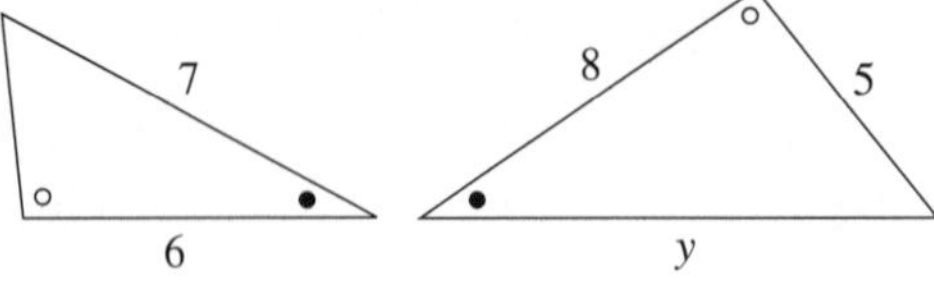

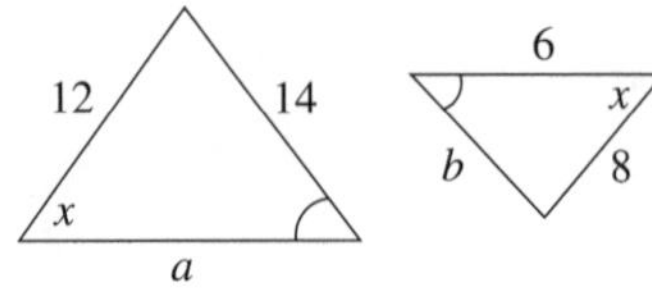

4 Which of these is the value of z?

(A) $3\frac{3}{4}$ (B) $4\frac{1}{4}$ (C) $4\frac{1}{2}$ (D) $4\frac{3}{4}$ 1

5 Which of these is the value of y?

(A) $9\frac{1}{3}$ (B) $9\frac{2}{3}$ (C) 10 (D) $10\frac{1}{3}$ 1

6 Which of these is the value of a?

(A) 9 (B) $9\frac{1}{2}$ (C) $9\frac{3}{4}$ (D) 10 1

7 Which of these is the value of b?

(A) $8\frac{2}{3}$ (B) $9\frac{1}{3}$ (C) $9\frac{2}{3}$ (D) $10\frac{1}{3}$ 1

The diagrams show similar triangles and are used to answer questions 8 to 13.

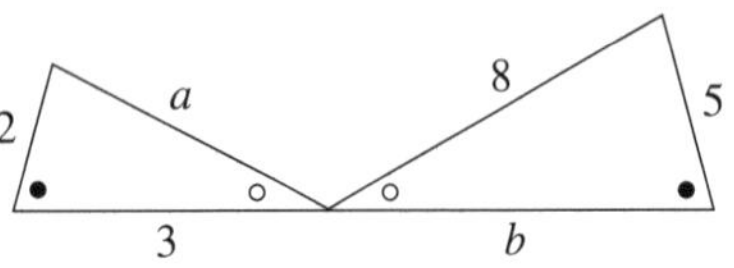

Figure 1

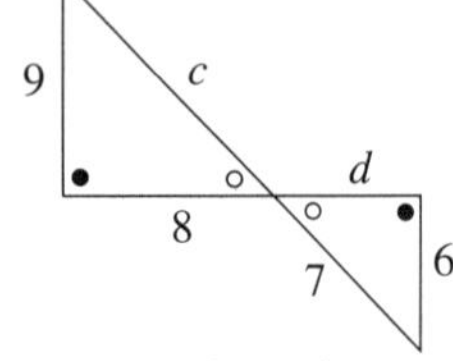

Figure 2

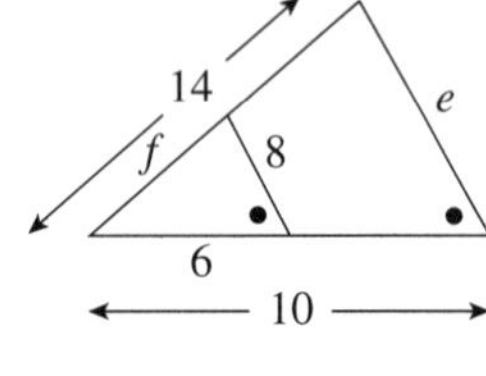

Figure 3

8 Using figure 1, which of these is the value of a?

(A) 3.2 (B) 3.4 (C) 3.6 (D) 3.8 1

9 Using figure 1, which of these is the value of b?

(A) 7 (B) 7.2 (C) 7.4 (D) 7.5 1

10 Using figure 2, which of these is the value of c?

(A) 10.3 (B) 10.4 (C) 10.5 (D) 10.6 1

11 Using figure 2, which of these is closest to the value of d?

(A) 5.2 (B) 5.3 (C) 5.4 (D) 5.5 1

12 Using figure 3, which of these is closest to the value of e?

(A) 13.3 (B) 13.4 (C) 13.5 (D) 13.6 1

13 Using figure 3, which of these is the value of f?

(A) 8 (B) 8.2 (C) 8.4 (D) 8.6 1

Similarity

LEVEL 2 TEST — PART B

Write answers and working in the space provided.

Marks

The diagrams are used to answer questions 14 and 15.

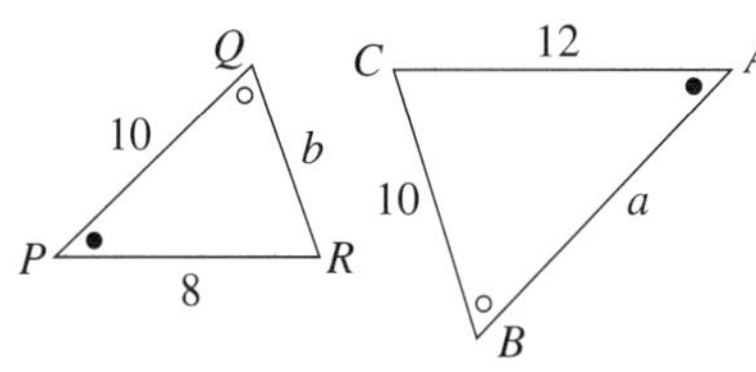

Figure 1

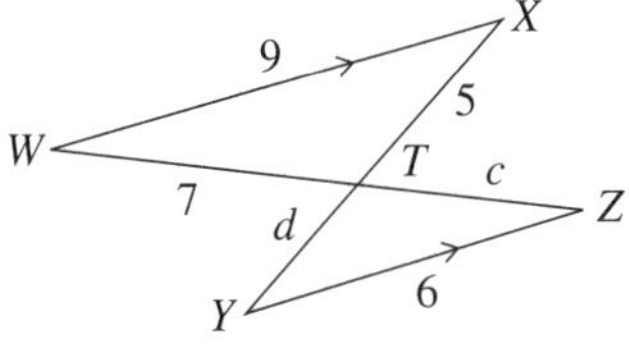

Figure 2

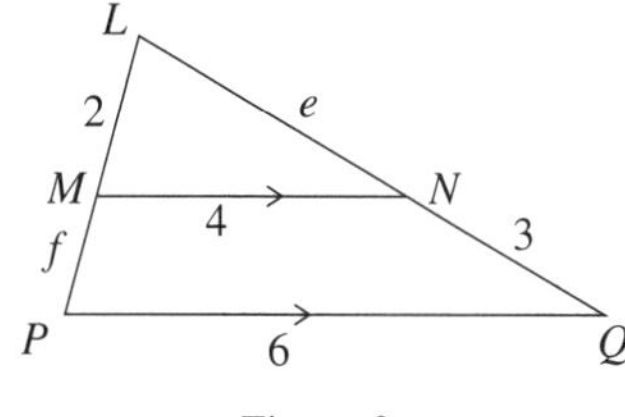

Figure 3

14 Complete the similarity statements for each pair of similar triangles, making sure the letters are in matching order:

a $\Delta PQR \parallel\!| \Delta$__________ **b** $\Delta WXT \parallel\!| \Delta$__________ **c** $\Delta LQP \parallel\!| \Delta$__________ 3

15 What is the value of:

a a? **b** b? **c** c?

6

d d? **e** e? **f** f?

6

The diagrams are used to answer questions 16 and 17.

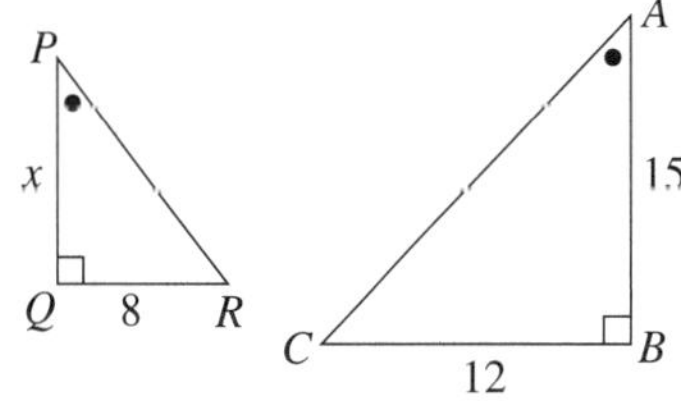

Figure 4

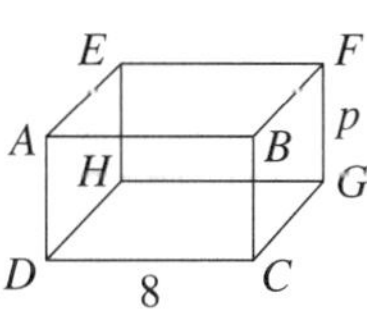

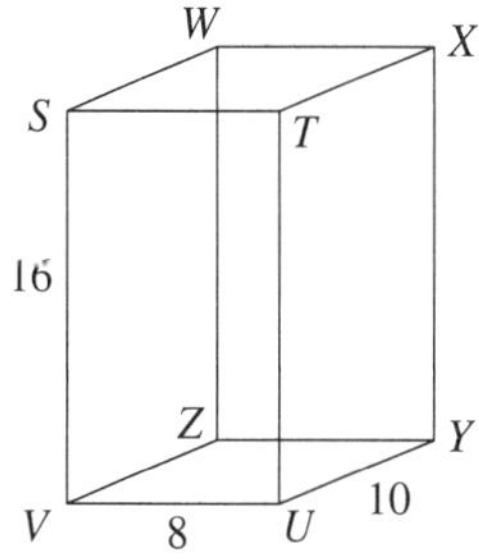

Figure 5

16 Using figure 4, what is the:

a value of x? **b** ratio of $PR:AC$? **c** ratio of the areas of the triangles?

6

17 In the rectangular prisms in figure 5, $\dfrac{ST}{BC} = \dfrac{SW}{BF} = \dfrac{SV}{BA}$. What is the:

a value of p? **b** ratio of the surface areas? **c** ratio of the volumes?

6

Total marks /40

Similarity

LEVEL 3 TEST — PART A

Time allowed: 40 minutes

Marks

A triangle is enlarged by a factor of 2.5 and the original triangle and its image are used to answer questions 1 to 3.

1 Which of these is the length of a side on the image if the matching side on the original triangle was 5 cm in length?

(A) 2 cm (B) 5 cm (C) 7.5 cm (D) 12.5 cm — 1

2 Which of these is the length of a side of the original triangle if the matching side on the image was 10 cm in length?

(A) 2 cm (B) 2.5 cm (C) 4 cm (D) 7.5 cm 

3 Which of these is the ratio of the area of the original triangle to the area of the image?

(A) 2 : 5 (B) 1 : 10 (C) 4 : 15 (D) 4 : 25 

The diagrams show three pairs of similar triangles and are used to answer questions 4 to 10.

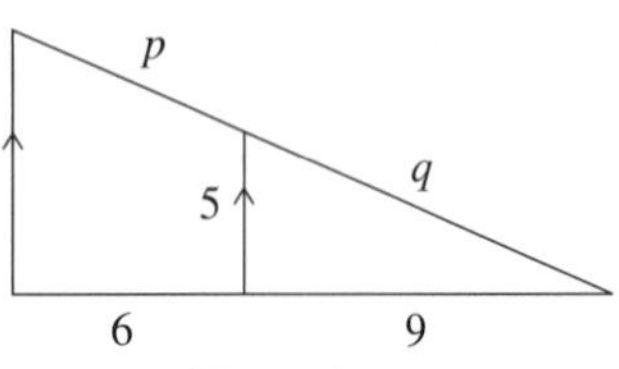

Figure 1

Figure 2

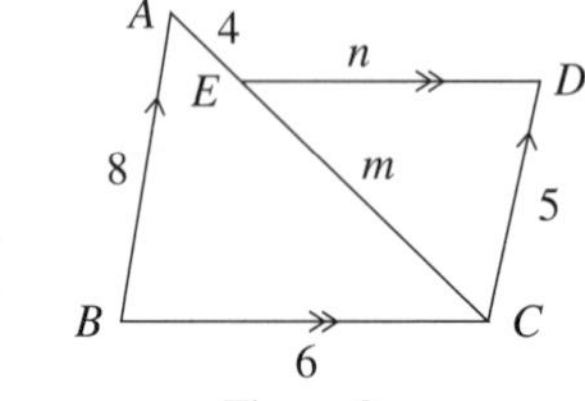

Figure 3

4 Using figure 1, which of these is the value of y?

(A) $\frac{25}{3}$ (B) $\frac{15}{2}$ (C) $\frac{13}{2}$ (D) $\frac{35}{3}$ — 1

5 Using figure 1, which of these is a correct equation?

(A) $\frac{p}{y} = \frac{q}{9}$ (B) $\frac{y}{5} = \frac{p}{q}$ (C) $\frac{q}{p} = \frac{p}{15}$ (D) $\frac{y}{p+q} = \frac{5}{q}$ — 1

6 Using figure 2, which of these is a correct equation?

(A) $5a = 2b$ (B) $2a = 5b$ (C) $ac = 10$ (D) $bc = 10$ — 1

7 Using figure 2, which of these is the value of c?

(A) 8 (B) 10 (C) 12 (D) 20 — 1

8 Using figure 3, which of these is the correct statement?

(A) $\Delta ABC ||| \Delta CDE$ (B) $\Delta ABC ||| \Delta DEC$ (C) $\Delta ABC ||| \Delta EDC$ (D) $\Delta ABC ||| \Delta DCE$ — 1

9 Using figure 3, which of these is a correct equation?

(A) $\frac{n}{8} = \frac{6}{5}$ (B) $\frac{n}{6} = \frac{5}{8}$ (C) $\frac{n}{6} = \frac{8}{5}$ (D) $\frac{n}{8} = \frac{5}{6}$ — 1

10 Using figure 3, which of these is a correct equation?

(A) $\frac{m+4}{m} = \frac{8}{5}$ (B) $\frac{m+4}{m} = \frac{5}{8}$ (C) $\frac{m+4}{n} = \frac{6}{8}$ (D) $\frac{m+4}{n} = \frac{8}{6}$ — 1

A rectangle with length 12 m and area 60 m^2 is enlarged by a factor of 3 and is used to answer questions 11 to 13.

11 Which of these is the width of the image?

(A) 15 m (B) 27 m (C) 36 m (D) 45 m — 1

12 What is the ratio of the perimeter of the image to the perimeter of the original rectangle?

(A) 1 : 3 (B) 3 : 1 (C) 1 : 9 (D) 9 : 1 — 1

13 The cost of covering the original rectangle is $36. What will be the cost of covering **half** of the image?

(A) $108 (B) $162 (C) $216 (D) $324 — 1

LEVEL 3 TEST

PART B

Write answers and working in the space provided.

Marks

14 Find the exact value of x:

a

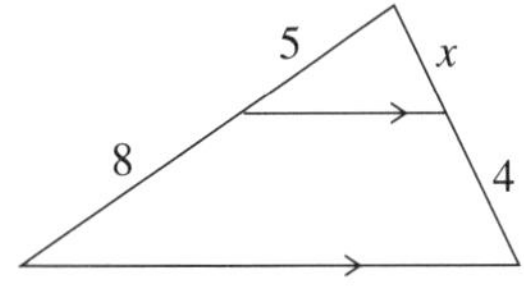

b

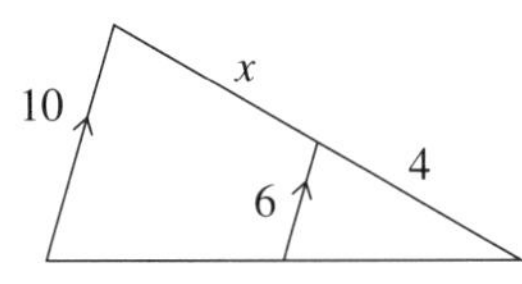

c

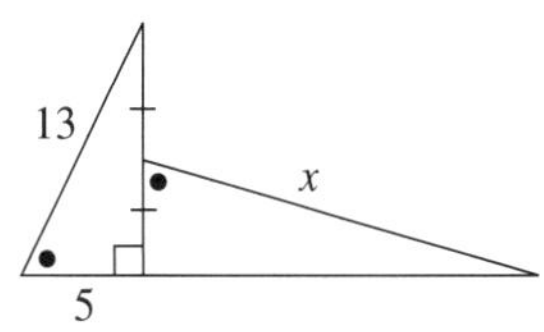

9

15 Find the value of a, and hence the values of b and c:

a

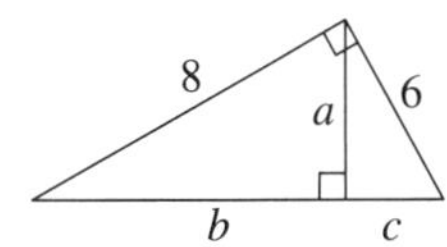

b

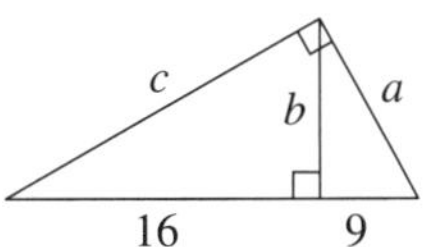

c

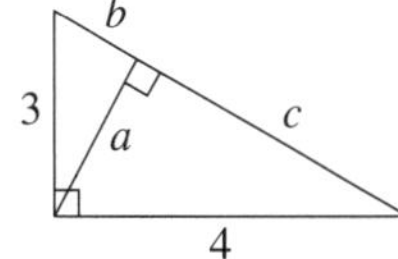

9

16 Solve the problems by first drawing a sketch.

a A triangle with side lengths 5 cm, 10 cm and 12 cm is similar to another triangle with shortest side 12 cm. What is the perimeter of the larger triangle?

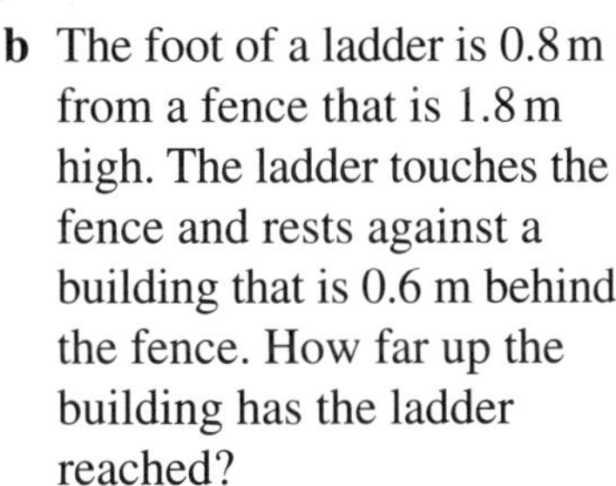

b The foot of a ladder is 0.8 m from a fence that is 1.8 m high. The ladder touches the fence and rests against a building that is 0.6 m behind the fence. How far up the building has the ladder reached?

c A person 1.8 m tall who is standing 6 m from a streetlight casts a 2.7 m shadow. If she moves 2 m further from the streetlight, what will be the length of her shadow?

Total marks

Measurement and Geometry Test: Level 1

STRAND TEST **PART A**

Time allowed: 50 minutes

Marks

The diagrams show triangles *ABC*, *LMN* and *PQR* and are used to answer questions 1 to 3.

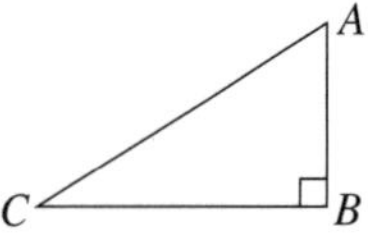

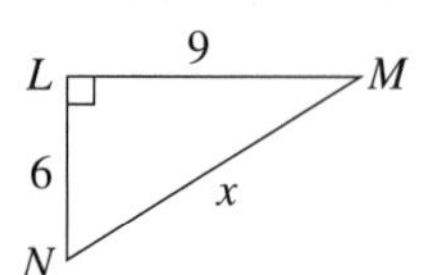

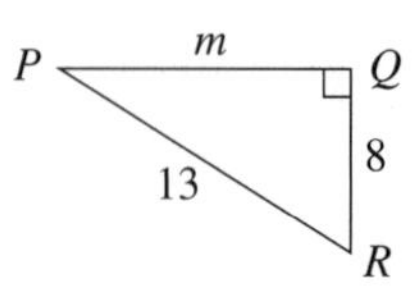

1 Using triangle *ABC*, which of the following is correct?

(A) $BC^2 = AC^2 \times AB^2$ (B) $AB^2 = BC^2 + AC^2$ (C) $BC^2 = AC^2 + AB^2$ (D) $AC^2 = AB^2 + BC^2$

2 Using triangle *LMN*, which of these shows the correct relationship?

(A) $x = 6^2 + 9^2$ (B) $x^2 = 6^2 + 9^2$ (C) $x^2 = 6^2 - 9^2$ (D) $x^2 = 9^2 - 6^2$

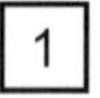

3 Using triangle *PQR*, which of these shows the correct relationship?

(A) $m^2 = 8^2 + 13^2$ (B) $m^2 = 13^2 \div 8^2$ (C) $m^2 = 13^2 - 8^2$ (D) $m^2 = 8^2 - 13^2$

4 Which of these is 91.4992 m rewritten to the nearest centimetre?

(A) 1 cm (B) 91 cm (C) 915 cm (D) 9150 cm

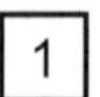

5 What is the surface area of a rectangular prism of dimensions 8 cm, 6 cm and 5 cm?

(A) 118 cm^2 (B) 236 cm^2 (C) 240 cm^2 (D) 256 cm^2

1

6 Which of these is closest to the volume of a cylinder with radius 9 cm and height 14 cm?

(A) 792 cm^3 (B) 890 cm^3 (C) 3407 cm^3 (D) 3563 cm^3

7 Which of these is the surface area of a cube with a side length of 5.4 cm?

(A) 32.4 cm^2 (B) 129.6 cm^2 (C) 157.64 cm^2 (D) 174.96 cm^2

8 A semicircle has a diameter of 16 cm. What is the area correct to two decimal places?

(A) 100.53 cm^2 (B) 201.06 cm^2 (C) 226.81 cm^2 (D) 402.12 cm^2

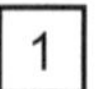

The diagrams are used to answer questions 9 to 12.

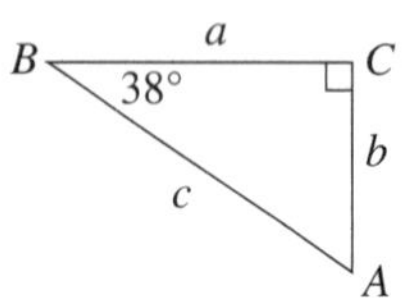

Figure 1

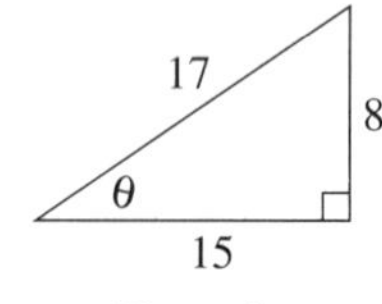

Figure 2

17
8
θ
15

Figure 3

9 In figure 1, which of these is the adjacent side to 38°?

(A) *a* (B) *b* (C) *c* (D) *AB*

10 In figure 1, which of these is equal to sin 38°?

(A) $\frac{a}{c}$ (B) $\frac{a}{b}$ (C) $\frac{b}{c}$ (D) $\frac{c}{b}$

11 In figure 2, which of these is equal to the expression for *m*?

(A) $m = 7 \times \sin 39°$ (B) $m = 7 \times \cos 39°$ (C) $m = 7 \times \tan 39°$ (D) $m = 39 \times \sin 7°$

12 In figure 3, which of these equations is used to find the size of angle θ?

(A) $\cos\theta = \frac{15}{17}$ (B) $\tan\theta = \frac{8}{17}$ (C) $\tan\theta = \frac{15}{17}$ (D) $\sin\theta = \frac{15}{17}$

1

13 A photograph measures 24 cm by 18 cm. The photograph is to be enlarged so that the new photo will be 27 cm wide. What is the new length?

(A) 30 cm (B) 32 cm (C) 34 cm (D) 36 cm

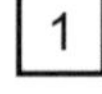

Measurement and Geometry Test: Level 1

STRAND TEST — PART B

Write answers and working in the space provided.

Marks

14 Find the value of the pronumeral, correct to two decimal places:

a

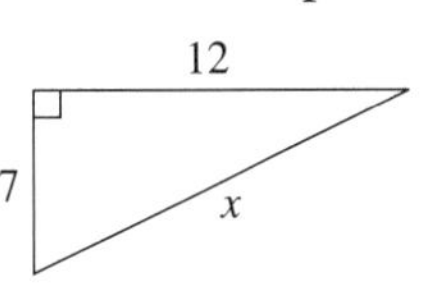

b

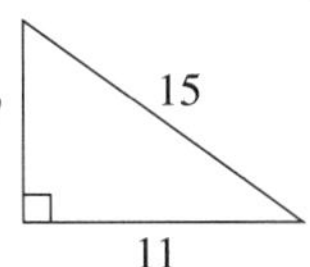

c

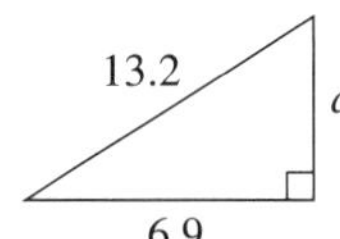

6

15 Determine whether a right triangle can have sides of 9 cm, 12 cm and 15 cm.

2

16 The diagram shows a triangular prism.

a Find the volume of the prism.

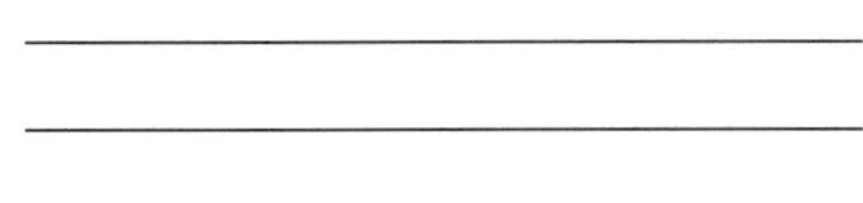

D
A
6 cm
F
E
7 cm
B
8 cm
C

2

b Use Pythagoras' theorem to find the length of *AC*.

c Find the surface area of the prism.

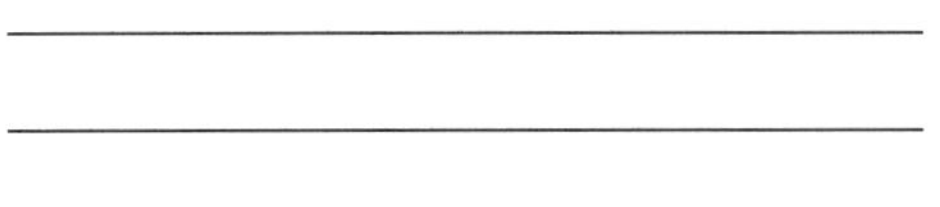

4

17 What is the volume of the solid?

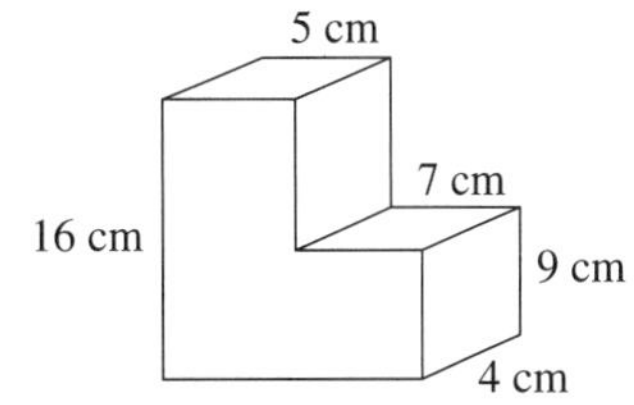

3

18 Find the value of the pronumeral in each of the triangles, correct to two decimal places:

a

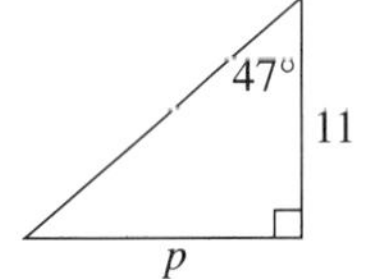

b

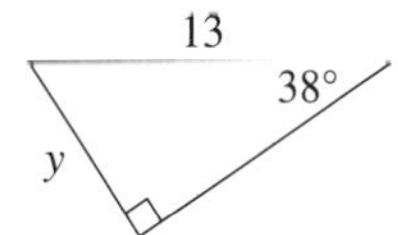

4

Measurement and Geometry Test: Level 1

STRAND TEST PART B

Write answers and working in the space provided.

Marks

19 Find the size of the angle θ, to the nearest degree:

a

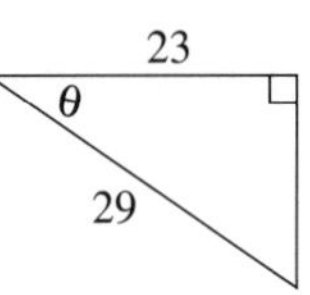

b

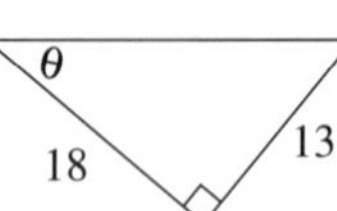

4

20 $ABCD$ is enlarged to produce $PQRS$.

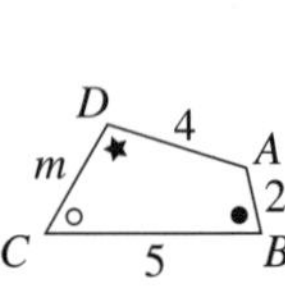

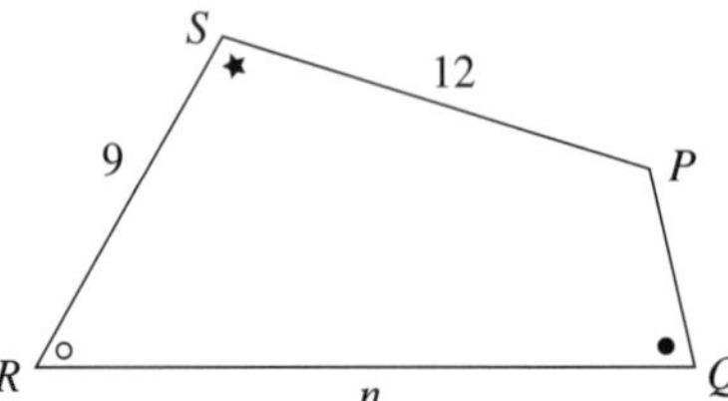

a What is the enlargement factor?

b What side matches RS?

c What angle matches DAB?

3

d Complete $\frac{SR}{DC} = \frac{\quad}{BC}$

e What is the value of m?

f What is the value of n?

3

21 Find the value of the pronumerals in the pair of similar shapes.

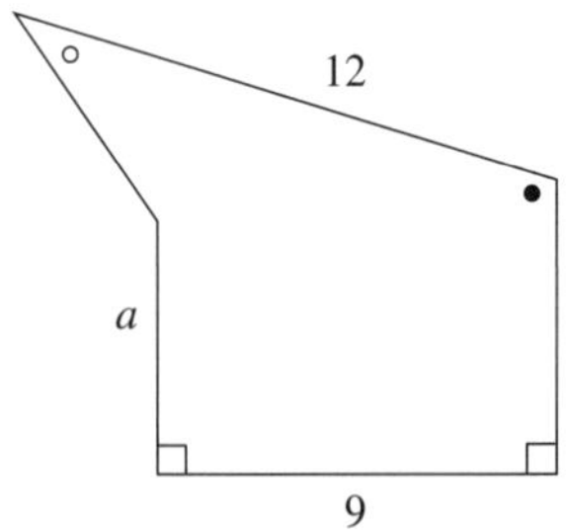

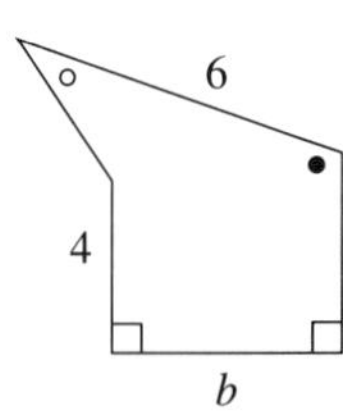

2

22 On a sunny afternoon Jack measured the shadow of a flagpole to be 6 m. If he knows the height of the pole is 4 m, what will be the length of the shadow of the following objects at that same time?

a a building 12 m high

b a tree 7.2 m tall

4

Total marks

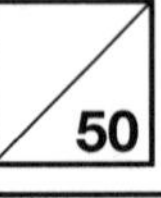

Measurement and Geometry Test: Level 2

STRAND TEST — PART A

Time allowed: 50 minutes

Marks

The diagrams show triangles *EFG*, *XYZ*, *ABC* and *LMN* and are used to answer questions 1 to 4.

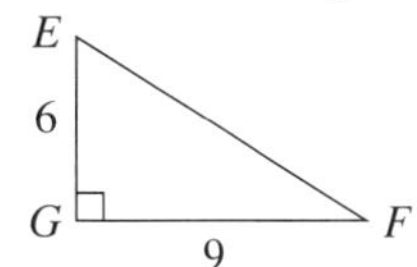

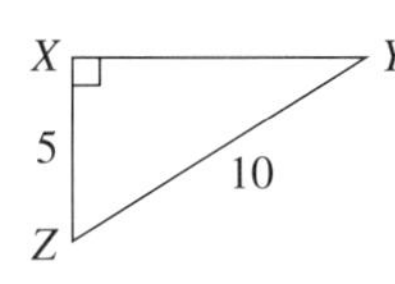

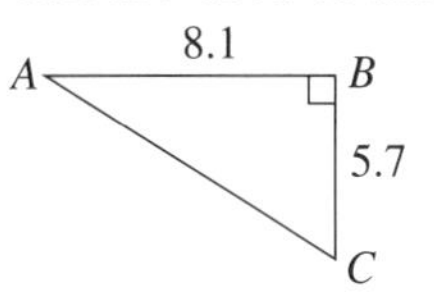

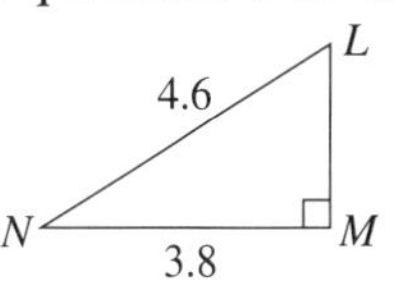

1 Using triangle *EFG*, which of these is closest to the length of *EF*?

Ⓐ 10.7 Ⓑ 10.8 Ⓒ 12 Ⓓ 15 [1]

2 Using triangle *XYZ*, which of these is closest to the length of *XY*?

Ⓐ 8.6 Ⓑ 8.7 Ⓒ 11.1 Ⓓ 11.2 [1]

3 Using triangle *ABC*, which of these is the length of *AC*, to three decimal places?

Ⓐ 9.900 Ⓑ 9.904 Ⓒ 9.905 Ⓓ 9.910 [1]

4 Using triangle *LMN*, which of these is the length of *LM*, to two significant figures?

Ⓐ 2.5 Ⓑ 2.59 Ⓒ 2.6 Ⓓ 2.60 [1]

5 Which of these is closest to the curved surface area of a cylinder of radius 3 cm and height 5 cm?

Ⓐ 93 cm^2 Ⓑ 94 cm^2 Ⓒ 95 cm^2 Ⓓ 96 cm^2 [1]

6 The volume of a cube is 64 cm^3. What is the surface area of the cube?

Ⓐ 48 cm^2 Ⓑ 84 cm^2 Ⓒ 96 cm^2 Ⓓ 192 cm^2 [1]

7 Which of these is the capacity of a cylinder with diameter 12 cm and height 15 cm, correct to three significant figures?

Ⓐ 1690 mL Ⓑ 1700 mL Ⓒ 6780 mL Ⓓ 6790 mL [1]

8 When θ is acute, which of these is closest to the value of θ when $\cos\theta = 0.7015$?

Ⓐ 45° Ⓑ 46° Ⓒ 47° Ⓓ 48° [1]

9 Which of these is closest to the value of θ if $\sin\theta = \frac{6}{13}$, given θ is acute?

Ⓐ 27°29′ Ⓑ 27°30′ Ⓒ 27°48′ Ⓓ 27°49′ [1]

The diagrams are used to answer questions 10 to 13.

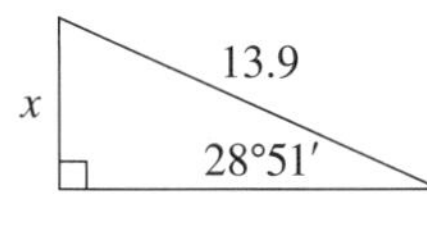

Figure 1

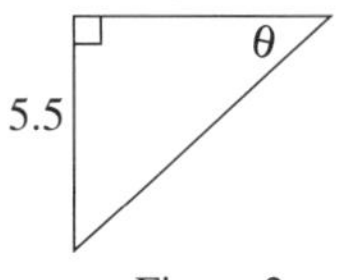

Figure 2

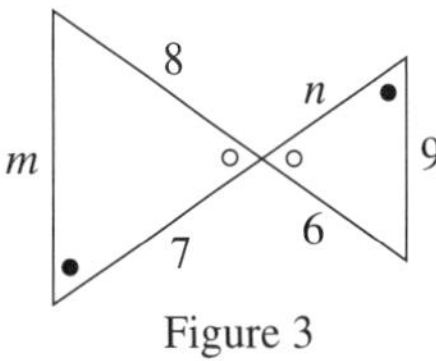

Figure 3

10 Using figure 1, what is the value of *x*, correct to two decimal places?

Ⓐ 6.63 Ⓑ 6.64 Ⓒ 6.70 Ⓓ 6.71 [1]

11 Using figure 2, what is the value of θ, to the nearest minute?

Ⓐ 15°20′ Ⓑ 43°24′ Ⓒ 46°36′ Ⓓ 46°37′ [1]

12 Using figure 3, which of these is the value of *m*?

Ⓐ 10 Ⓑ 10.5 Ⓒ 12 Ⓓ 12.5 [1]

13 Using figure 3, which of these is closest to the value of *n*?

Ⓐ 5.3 Ⓑ 5.8 Ⓒ 6.4 Ⓓ 6.9 [1]

Measurement and Geometry Test: Level 2

STRAND TEST PART B

Write answers and working in the space provided.

Marks

14 Find the value of the pronumeral, correct to three significant figures:

a

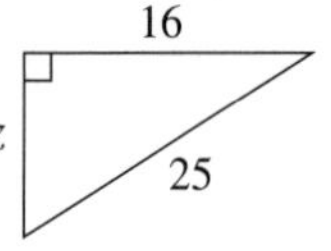

b

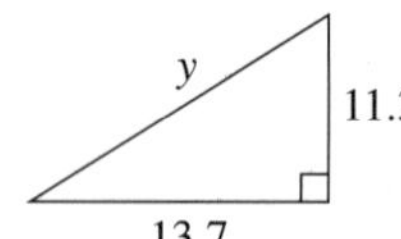

4

15 Two poles are positioned 3 m apart. The poles have heights of 2.9 m and 3.7 m. Find the length of a wire that links the tops of the two poles, correct to two decimal places.

3

16

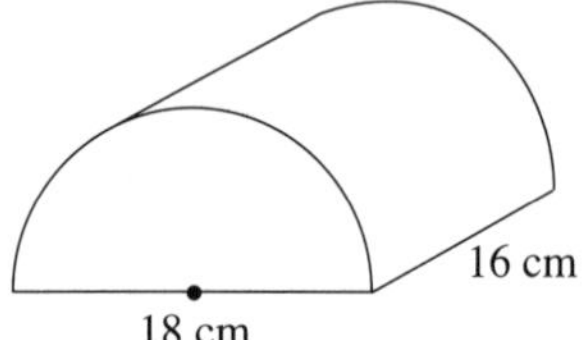

a Find the volume, correct to three decimal places.

3

b Find the surface area, correct to three decimal places.

3

17

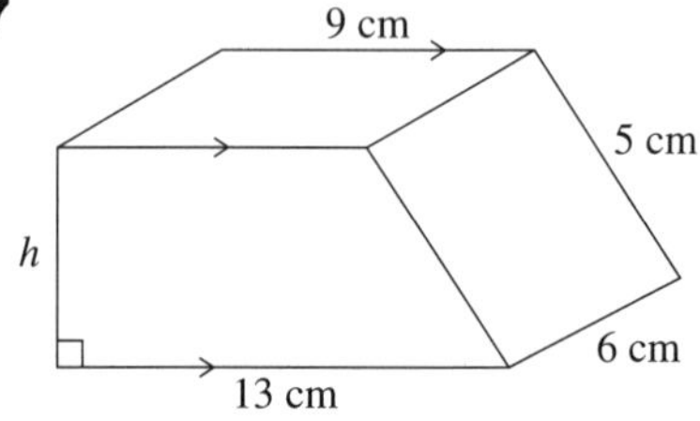

a Find the value of h.

2

b Find the surface area, correct to two significant figures.

3

18 If $\cos\,\theta = \frac{6}{10}$, and θ is acute, what is the value of $\tan\,\theta$?

2

19 The diagonal of a rectangle is 12.8 cm in length and makes an angle of $36°11'$ with the longer side. What is the length of the rectangle, correct to two decimal places?

2

Measurement and Geometry Test: Level 2

STRAND TEST — PART B

Write answers and working in the space provided.

Marks

20

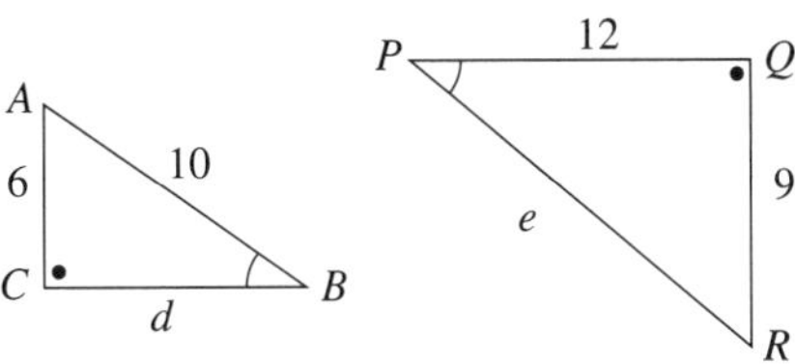

a Complete the similarity statement for the pair of triangles, making sure the letters are in matching order: ΔBCA ||| Δ ______________ 1

b Find the value of:

i d ______________ **ii** e ______________ 2

c Find the ratio of the area of the two triangles. ______________ 2

21

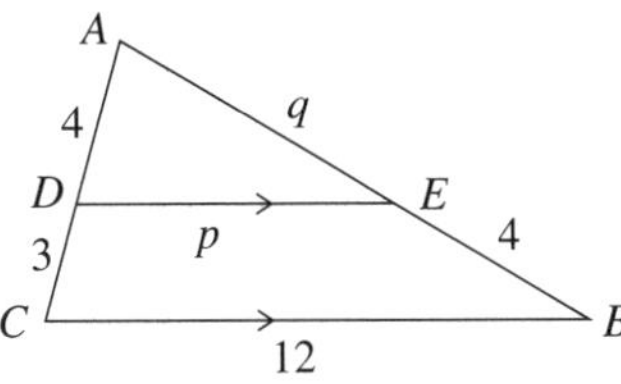

Find the value of:

a p ______________ **b** q ______________ 4

22 Find the value of x.

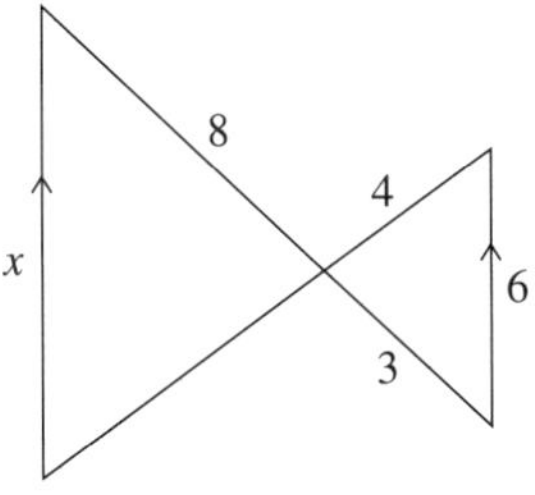

2

23 The diagram shows two similar cylinders. If the larger cylinder has a height of 15 cm, find the:

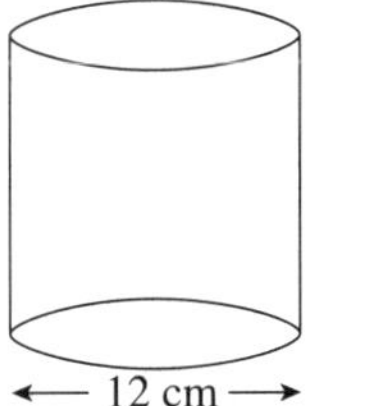

a height of the smaller cylinder ______________ 1

b ratio of their heights ______________ 1

c ratio of their surface areas ______________ 2

Total marks

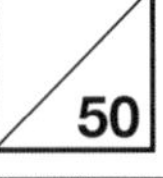

Measurement and Geometry Test: Level 3

STRAND TEST PART A

Time allowed: 50 minutes

Marks

1 What is the length of the shortest side of a right triangle with two sides of $\sqrt{48}$ cm and 8 cm?

(A) 4 cm (B) 6 cm (C) 8 cm (D) $\sqrt{112}$ cm 1

2 A yacht rounds a channel marker and sails 340 m west and then 160 m south. Which of these is the distance, to the nearest metre, that the yacht is from the marker?

(A) 376 m (B) 377 m (C) 378 m (D) 379 m 1

3 The area of a square is 8 cm². Which of these is the length of its diagonal?

(A) $\sqrt{12}$ cm (B) 4 cm (C) $\sqrt{18}$ cm (D) $\sqrt{24}$ cm 1

4 A closed cylinder has a height of 8 cm and a diameter of 6 cm. Which of these is the surface area of the cylinder?

(A) 66π cm² (B) 72π cm² (C) 80π cm² (D) 114π cm² 1

5 A rectangular prism with a volume of 192 cm³ has a length of 8 cm and height of 6 cm. Which of these is the surface area of the prism?

(A) 208 cm² (B) 228 cm² (C) 256 cm² (D) 324 cm² 1

6 The curved surface area of a cylinder with diameter 16 cm is 192π cm². Which of these is the capacity of the cylinder?

(A) 624π mL (B) 768π mL (C) 1448π mL (D) 1536π mL 1

7 What is the maximum circumference of a circle that can fit inside a square with perimeter 64 cm?

(A) 4π cm (B) 8π cm (C) 16π cm (D) 22π cm 1

The diagrams are used to answer questions 8 to 13.

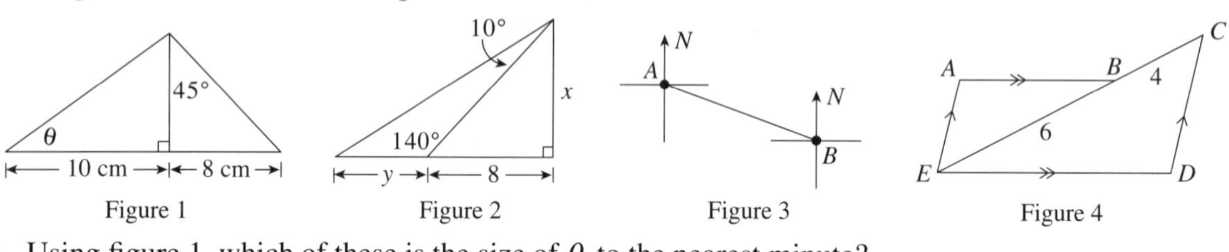

Figure 1 Figure 2 Figure 3 Figure 4

8 Using figure 1, which of these is the size of θ, to the nearest minute?

(A) 38°38′ (B) 38°39′ (C) 38°40′ (D) 38°41′ 1

9 Using figure 2, which of these is closest to the value of x?

(A) 5.1 (B) 6.1 (C) 6.7 (D) 9.5 1

10 Using figure 2, which of these is the value of y, correct to three significant figures?

(A) 2.48 (B) 3.63 (C) 10.4 (D) 11.6 1

11 Using figure 3, if the bearing of A from B is 315°, what is the bearing of B from A?

(A) 125° (B) 135° (C) 145° (D) 155° 1

12 Using figure 4, if $CD = 8$ cm, which of these is closest to the length of AE?

(A) 4.8 (B) 5 (C) 5.2 (D) 5.4 1

13 Using figure 4, if $AB = 5$ cm, which of these is closest to the length of DE?

(A) 7.8 (B) 8.2 (C) 8.3 (D) 8.5 1

Measurement and Geometry Test: Level 3

STRAND TEST — PART B

Write answers and working in the space provided.

Marks

14 The three sides of a right triangle are $(3a + 1)$ cm, $3a$ cm and $(a + 1)$ cm. Find the value of a.

3

15 A plane left X and flew south-east for 28 km to Y. How far, to the nearest kilometre, is the plane east of X?

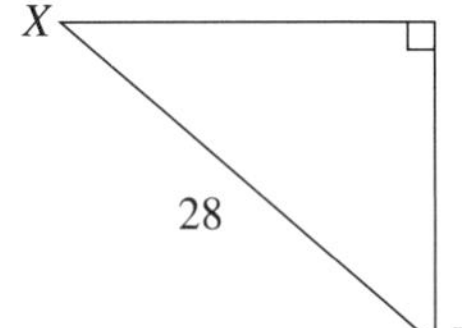

3

16 The diagram shows a rectangular prism where $BC = 8$ cm, $AG = 12$ cm and the area of $BCGF$ is 40 cm^2.

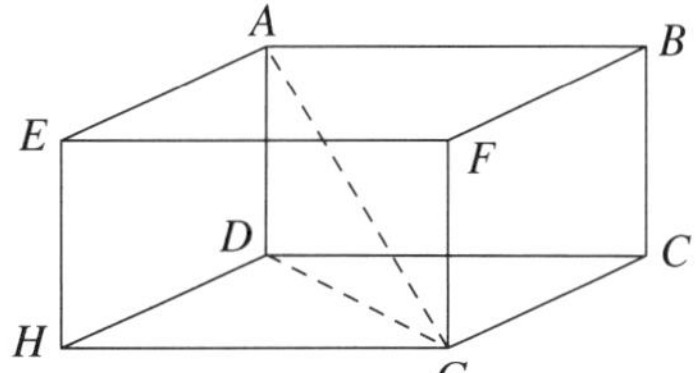

a Find the exact length of DG.

2

b Find the exact length of HG.

c Find the volume of the prism, correct to three significant figures.

3

17 Theon is building two shade-houses for his garden plants using plastic sheeting. Using calculations, compare the:

a heights of the two structures

3

b volumes of the two structures

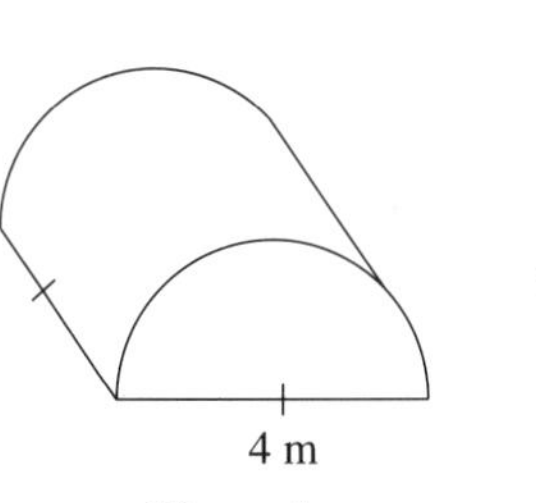
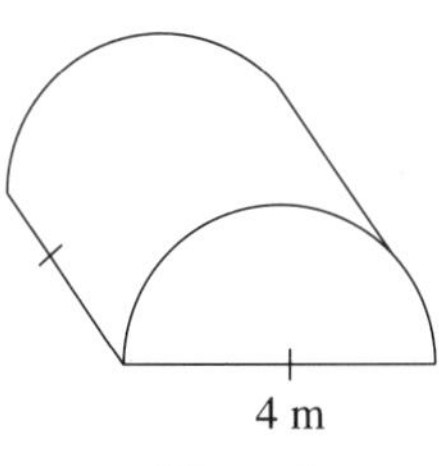

Figure A Figure B

3

18 Find the size of θ, to the nearest minute.

16
θ
52°
12

3

Measurement and Geometry Test: Level 3

STRAND TEST — PART B

Write answers and working in the space provided.

Marks

19 Find an expression for the area of a trapezium with three sides a cm and its base $2a$ cm. [2]

20 A boat leaves X and sails 16 km on a bearing of 235° to Y. How far west of the starting point is the boat? Give your answer to the nearest kilometre. [3]

21 A right triangle has a height of 6 cm and a hypotenuse of 10 cm. The triangle is to be enlarged so that its area is 216 cm^2. What is the perimeter of the enlarged triangle? [2]

22

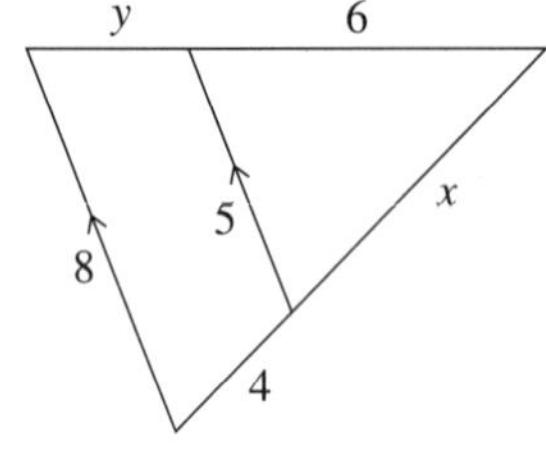

Find the value of:

a x

[2]

b y [2]

23

p cm
Y
Z
U
V
W
2.4 cm
X

a Explain why $\Delta XYZ \,|||\, \Delta UWV$. [2]

b Find the value of p.

c If the area of ΔUWV is 4.8 cm^2, what is the area of ΔXYZ? [4]

Total marks /50

Statistics

PRETEST

LEVEL 1

Time allowed: 10 minutes **Total marks: 10**

Marks

1 Which of these is the measure of location that represents the middle score in a data set?

Ⓐ mean Ⓑ median Ⓒ mode Ⓓ range [1]

2 Which of these is the score that occurs the most often in a data set?

Ⓐ mean Ⓑ median Ⓒ mode Ⓓ range [1]

The scores 19, 14, 16, 13, 13 are used to answer questions 3 to 6.

3 Which of these is the range of the scores?

Ⓐ 5 Ⓑ 6 Ⓒ 13 Ⓓ 16 [1]

4 Which of these is the mean of the scores?

Ⓐ 13 Ⓑ 15 Ⓒ 16 Ⓓ 64.6 [1]

5 Which of these is the mode of the scores?

Ⓐ 13 Ⓑ 15 Ⓒ 16 Ⓓ 19 [1]

6 Which of these is the median of the scores?

Ⓐ 13 Ⓑ 14 Ⓒ 15 Ⓓ 16 [1]

The frequency table below is used to answer questions 7 to 10.

Score (x)	Frequency (f)	$f \times x$
12	3	
13	5	
14	a	56
Total	12	b

7 Which of these is the range of scores?

Ⓐ 2 Ⓑ 3 Ⓒ 12 Ⓓ 29 [1]

8 Which of these is the value of a?

Ⓐ $a = 3$ Ⓑ $a = 4$ Ⓒ $a = 35$ Ⓓ $a = 42$ [1]

9 After completing the table, which of these is the value of b?

Ⓐ $b - 89$ Ⓑ $b = 146$ Ⓒ $b = 157$ Ⓓ $b = 168$ [1]

10 Which of these is closest to the mean of the scores?

Ⓐ 13.08 Ⓑ 13.15 Ⓒ 13.25 Ⓓ 13.65 [1]

Total marks /10

Statistics

PRETEST

LEVELS 2 AND 3

Time allowed: 10 minutes **Total marks: 10**

Marks

These data displays are used to answer questions 1 to 10.

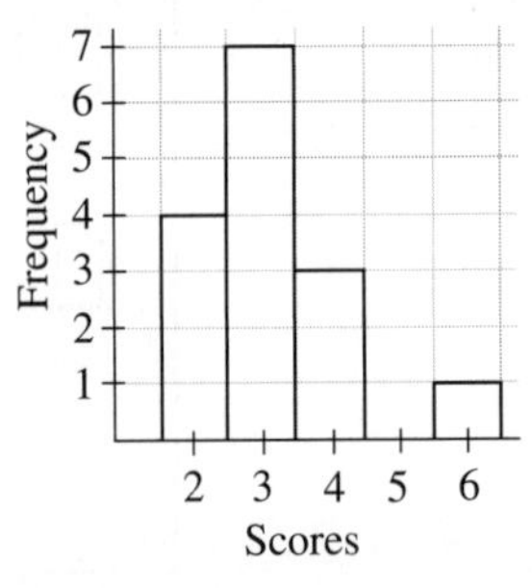

Figure 1

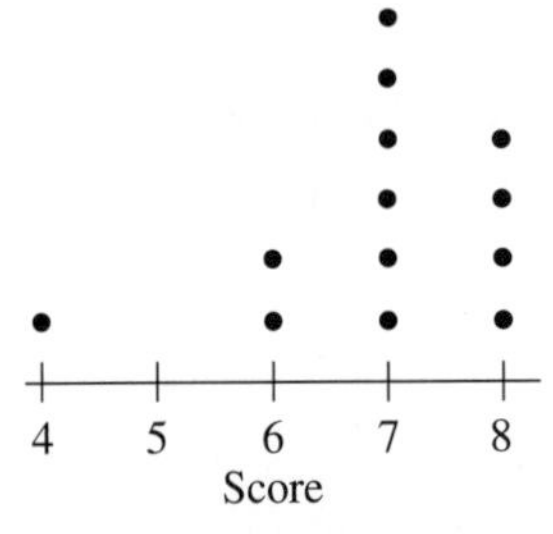

Figure 2

Stem	Leaf
3	1
4	89
5	25779
6	33388
7	2

Figure 3

1 Using figure 1, what is the mode of the scores?

(A) 3 (B) 4 (C) 5 (D) 7 — 1

2 Using figure 1, what is the median of the scores?

(A) 2 (B) 3 (C) 4 (D) 6 — 1

3 Using figure 1, what is the mean of the scores, correct to two decimal places?

(A) 3.00 (B) 3.03 (C) 3.13 (D) 3.75 — 1

4 Using figure 2, what is the range of the scores?

(A) 3 (B) 4 (C) 7 (D) 8 — 1

5 Using figure 2, what is the median of the scores?

(A) 3 (B) 4 (C) 6.5 (D) 7 — 1

6 Using figure 2, what is the mean of the scores, correct to three significant figures?

(A) 6.86 (B) 6.87 (C) 6.92 (D) 7.24 — 1

7 Using figure 3, what is the range of the scores?

(A) 1 (B) 31 (C) 41 (D) 63 — 1

8 Using figure 3, which of these is the outlier of the scores?

(A) 31 (B) 41 (C) 63 (D) 72 — 1

9 Using figure 3, which of these is the mean of the scores, correct to two significant figures?

(A) 56 (B) 57 (C) 58 (D) 60 — 1

10 Using figure 3, which of these is the median of the scores?

(A) 56.5 (B) 57 (C) 57.5 (D) 58 — 1

Total marks

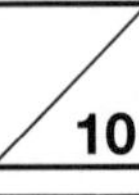

/10

Statistics

LEVEL 1 TEST — PART A

Time allowed: 40 minutes

Marks

The tables are used to answer questions 1 to 10.

Scores	Frequency	Cum. freq.
14	4	4
15	6	10
16	7	*b*
17	*a*	*c*
Total	20	

Figure 1

Stem	Leaf
6	348
7	2*e*49
8	012*d*

Figure 2

Class	Class	Frequency
16–24	20	*g*
25–33	29	8
34–42	*f*	11
43–51	47	5
Total		30

Figure 3

1 Using figure 1, which of these is the value of *a*?
(A) 3 (B) 5 (C) 7 (D) 19 — 1

2 Using figure 1, which of these is the value of *b*?
(A) 7 (B) 12 (C) 17 (D) 20 — 1

3 Using figure 1, which of these is the value of *c*?
(A) 6 (B) 10 (C) 14 (D) 20 — 1

4 Using figure 1, which of these is the mode?
(A) 10 (B) 15 (C) 16 (D) 20 — 1

5 Using figure 2, if the range is 24 what is the value of *d*?
(A) 2 (B) 4 (C) 7 (D) 9 — 1

6 Using figure 2, if the mode is 74 what is the value of *e*?
(A) 2 (B) 3 (C) 4 (D) not enough information — 1

7 Using figure 2, which of these is the median?
(A) 74 (B) 74.5 (C) 75 (D) 75.5 — 1

8 Using figure 3, which of these is the value of *f*?
(A) 35 (B) 36 (C) 37 (D) 38 — 1

9 Using figure 3, which of these is the value of *g*?
(A) 5 (B) 6 (C) 7 (D) 8 — 1

10 Using figure 3, which of these is the modal class?
(A) 16–24 (B) 25–33 (C) 34–42 (D) 43–51 — 1

Use the scores 12 15 18 19 21 23 30 to answer questions 11 to 13.

11 Which of these is the median of the scores?
(A) 18 (B) 19 (C) 20 (D) 30 — 1

12 Which of these is the lower quartile of the scores?
(A) 12 (B) 15 (C) 16.5 (D) 19 — 1

13 Which of these is the interquartile range of the scores?
(A) 4 (B) 7 (C) 8 (D) 18 — 1

LEVEL 1 TEST

PART B

Write answers and working in the space provided.

Marks

The statistical displays are used to answer questions 14 to 17.

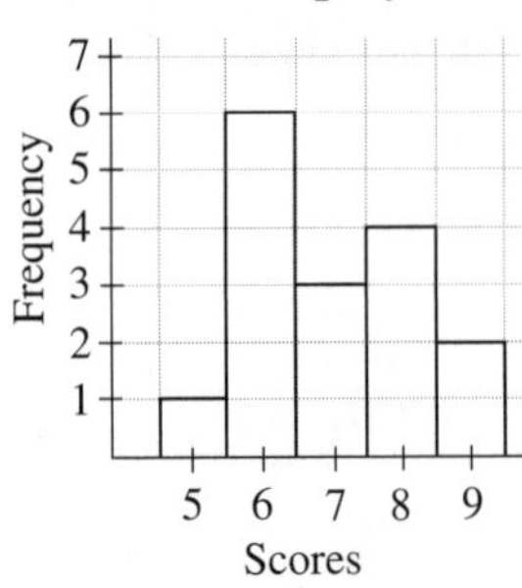

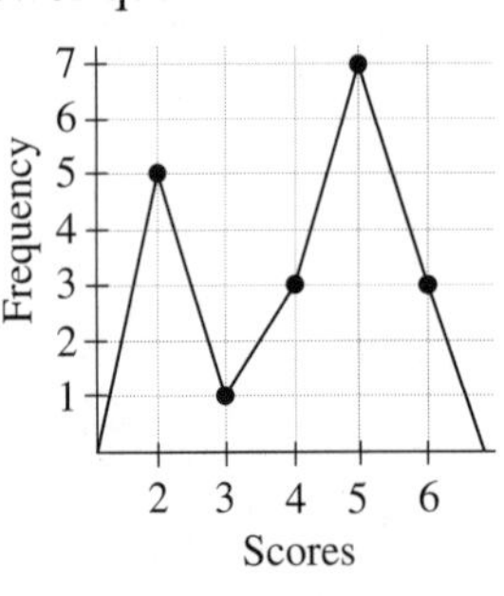

9M5		9M6
9	6	2558
8763	7	379
600	8	14

14 Using the histogram,

a find the mode ______

b find the range ______

c find the median ______ 3

d complete the table

Scores	Freq. (f)	$f \times x$
5		
6		
Total		

e calculate the mean ______ 4

15 Using the polygon,

a find the mode ______

b find the range ______

c find the median ______ 3

d complete the table

Scores	Freq. (f)	$f \times x$
2		
3		
Total		

e calculate the mean, correct to two decimal places ______ 6

16 The back-to-back stem-and-leaf display shows test results from two classes. Find the:

a number of students in 9M5 ______

b median mark in 9M6 ______

c mode of 9M5 ______ 3

d range of marks in 9M6 ______

e highest mark in 9M5 ______

f range of marks across the two classes ______ 3

17 For the scores 16, 18, 21, 22, 25, 27, 31, 33, 40, find the 5-point summary:

a minimum score ______

b lower quartile ______

c median ______ 3

d upper quartile ______

e maximum score ______ 2

Total marks

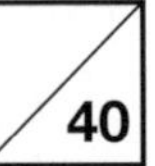

Statistics

LEVEL 2 TEST — PART A

Time allowed: 40 minutes

Marks

1 Which of these is a measure of spread?

(A) range (B) median (C) mean (D) mode 1

2 Consider the scores: 4, 6, 8, 10, 22. If the outlier is ignored, how has the mean changed?

(A) dropped by 1 (B) dropped by 3 (C) dropped by 4 (D) dropped by 12 1

The box plots are used to answer questions 3 to 8.

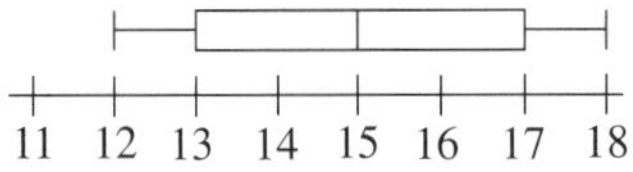

Figure 1

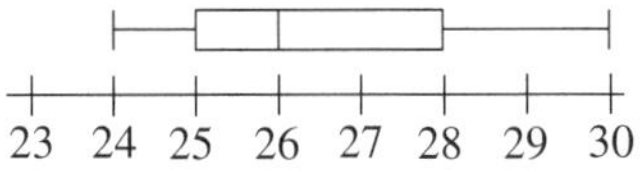

Figure 2

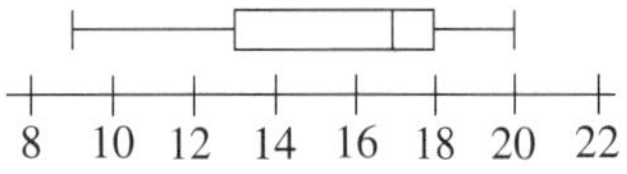

Figure 3

3 Using figure 1, which of these is the range?

(A) 4 (B) 6 (C) 7 (D) 15 1

4 Using figure 1, what percentage of the scores are from 13 to 18?

(A) 5% (B) 25% (C) 50% (D) 75% 1

5 Using figure 2, between which two scores lie 25% of the scores?

(A) 24 to 26 (B) 25 to 26 (C) 25 to 28 (D) 26 to 30 1

6 Using figure 2, what fraction of scores lie between 24 and 28?

(A) one-quarter (B) one-half (C) four-sevenths (D) three-quarters 1

7 Using figure 3, which of these is the value of the lower quartile?

(A) 9 (B) 12.5 (C) 13 (D) 17 1

8 Using figure 3, what is the difference between the maximum score and the median?

(A) 3 (B) 3.5 (C) 4 (D) 5 1

The diagrams are used to answer questions 9 to 13.

Scores	Freq.	Cum. freq.
34	6	6
35	14	20
36	11	31
37	9	40
Total	40	

Figure 4

Mass in kg

Girls		Boys
852	4	
9641	5	0579
840	6	14478
	7	39

Figure 5

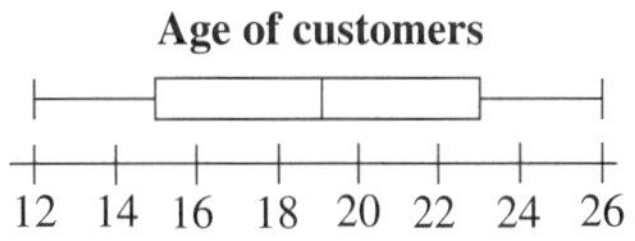

Age of customers

Figure 6

9 Using figure 4, what percentage of the scores were less than 36?

(A) 20% (B) 26% (C) 50% (D) 65% 1

10 Using figure 5, what is the difference between the medians of the girls and the boys?

(A) 6 (B) 7 (C) 8 (D) 9 1

11 Using figure 5, which of these is closest to the mean mass of the boys?

(A) 63.3 kg (B) 63.4 kg (C) 63.5 kg (D) 63.6 kg 1

12 Using figure 5, which of these is the overall median of the group?

(A) 57.5 (B) 58 (C) 58.5 (D) 59 1

13 Using figure 6, if there were 36 customers, how many were 15 years old and over?

(A) 21 (B) 24 (C) 27 (D) 75 1

Statistics

LEVEL 2 TEST — PART B

Write answers and working in the space provided.

Marks

The diagrams are used to answer questions 14 and 15.

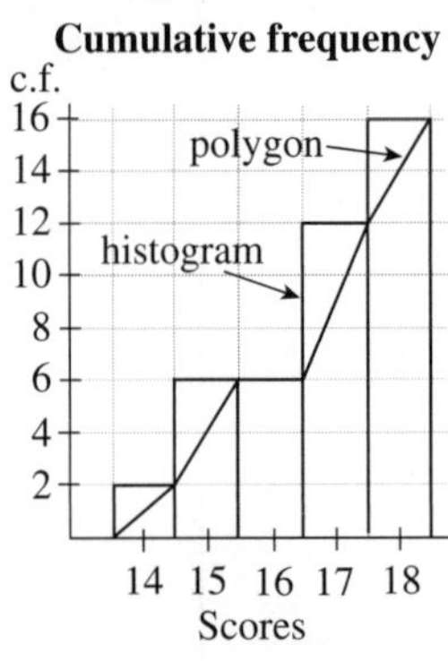

Figure 1

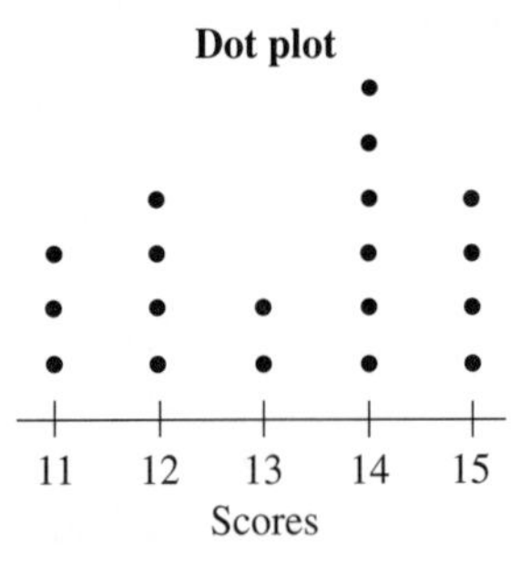

Figure 2

Grouped data table

Class	Class centre (x)	Frequency (f)	fx
0–6		3	
7–13		5	
14–20		7	
21–27		6	
28–34		9	
Totals			

Figure 3

14 Using figure 1, find the:

a range ______ **b** mode ______ **c** median ______ 3

d lower quartile ______ **e** upper quartile ______ **f** interquartile range ______ 3

15 Using figure 2, find the:

a range ______ **b** mode ______ **c** median ______ 3

d lower quartile ______ **e** upper quartile ______ **f** interquartile range ______ 3

16 Using figure 3:

a complete the table

b find the mean, correct to two decimal places ______

c what percentage of scores were less than 28? ______ 6

17 Complete 5-point summaries and draw box plots for the following:

a 16, 13, 18, 16, 19, 20, 14

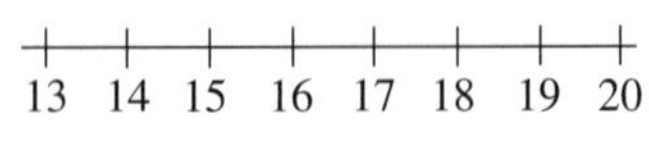

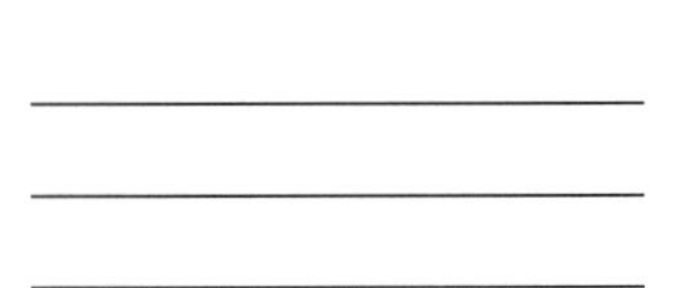

b 3, 6, 7, 4, 5, 7, 3, 8, 2, 9

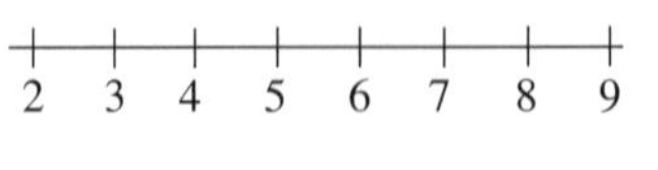

c 8, 10, 5, 7, 5, 9

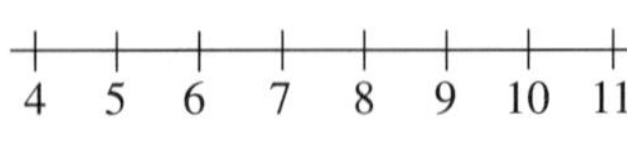

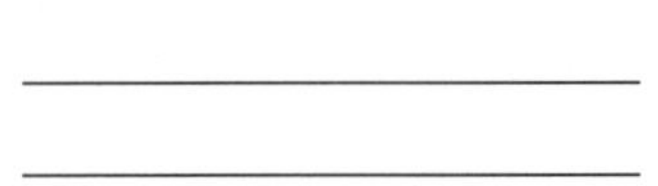

9

Total marks

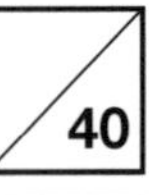

Statistics

LEVEL 3 TEST — PART A

Time allowed: 40 minutes

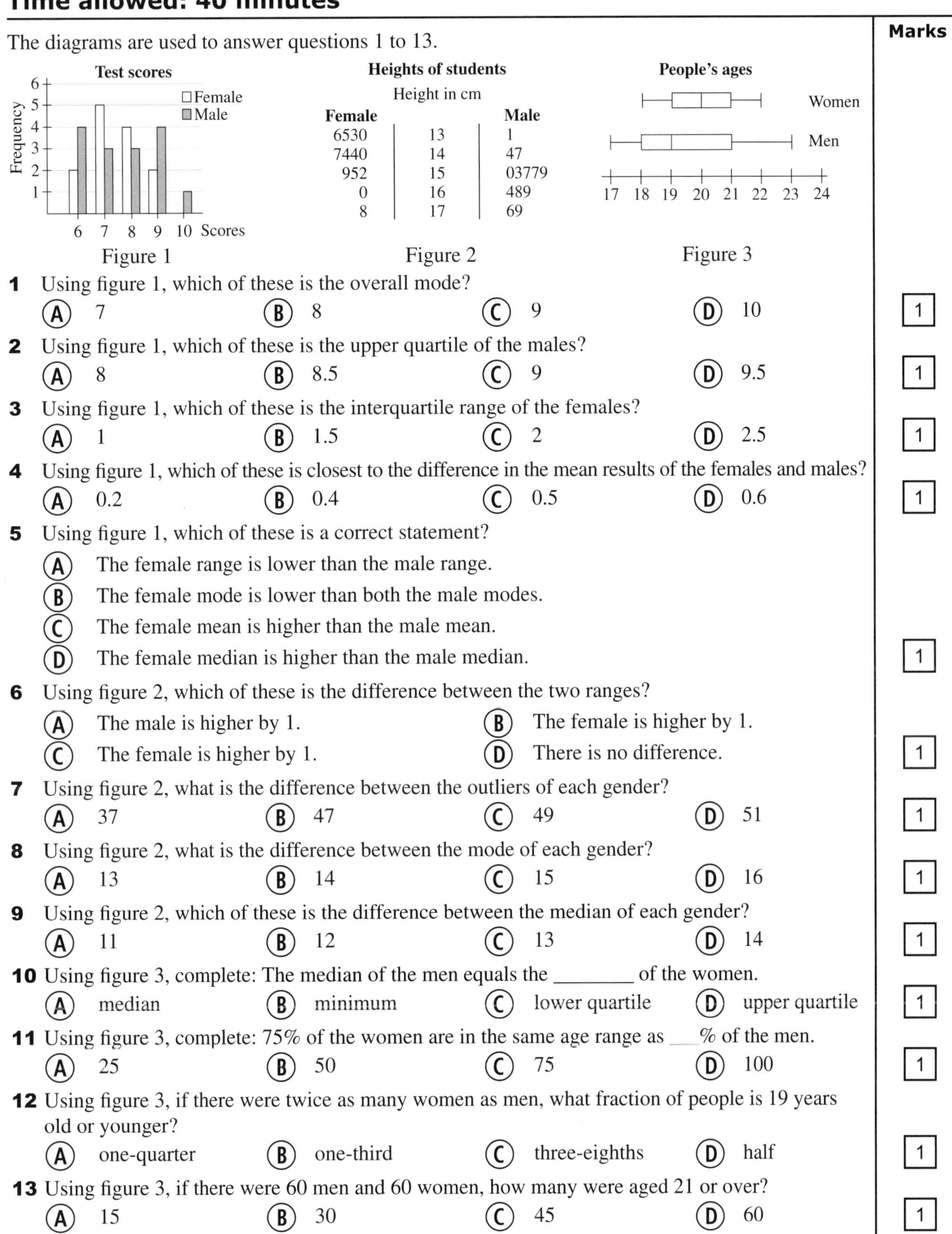

Marks

The diagrams are used to answer questions 1 to 13.

Figure 1

Heights of students

Height in cm

Female		Male
6530	13	1
7440	14	47
952	15	03779
0	16	489
8	17	69

Figure 2

Figure 3

1 Using figure 1, which of these is the overall mode?
(A) 7 (B) 8 (C) 9 (D) 10 — 1

2 Using figure 1, which of these is the upper quartile of the males?
(A) 8 (B) 8.5 (C) 9 (D) 9.5 — 1

3 Using figure 1, which of these is the interquartile range of the females?
(A) 1 (B) 1.5 (C) 2 (D) 2.5 — 1

4 Using figure 1, which of these is closest to the difference in the mean results of the females and males?
(A) 0.2 (B) 0.4 (C) 0.5 (D) 0.6 — 1

5 Using figure 1, which of these is a correct statement?
(A) The female range is lower than the male range.
(B) The female mode is lower than both the male modes.
(C) The female mean is higher than the male mean.
(D) The female median is higher than the male median. — 1

6 Using figure 2, which of these is the difference between the two ranges?
(A) The male is higher by 1. (B) The female is higher by 1.
(C) The female is higher by 1. (D) There is no difference. — 1

7 Using figure 2, what is the difference between the outliers of each gender?
(A) 37 (B) 47 (C) 49 (D) 51 — 1

8 Using figure 2, what is the difference between the mode of each gender?
(A) 13 (B) 14 (C) 15 (D) 16 — 1

9 Using figure 2, which of these is the difference between the median of each gender?
(A) 11 (B) 12 (C) 13 (D) 14 — 1

10 Using figure 3, complete: The median of the men equals the ________ of the women.
(A) median (B) minimum (C) lower quartile (D) upper quartile — 1

11 Using figure 3, complete: 75% of the women are in the same age range as ___% of the men.
(A) 25 (B) 50 (C) 75 (D) 100 — 1

12 Using figure 3, if there were twice as many women as men, what fraction of people is 19 years old or younger?
(A) one-quarter (B) one-third (C) three-eighths (D) half — 1

13 Using figure 3, if there were 60 men and 60 women, how many were aged 21 or over?
(A) 15 (B) 30 (C) 45 (D) 60 — 1

Statistics

LEVEL 3 TEST

PART B

Write answers and working in the space provided.

Marks

14 Consider the scores 12, 15, 13, 18, 40, 16, 15.

a Which score is the outlier? _______________ 1

b If the outlier is to be excluded, comment on the change to the:

i median **ii** mean **iii** interquartile range

6

The box plots are used to answer questions 15 and 16.

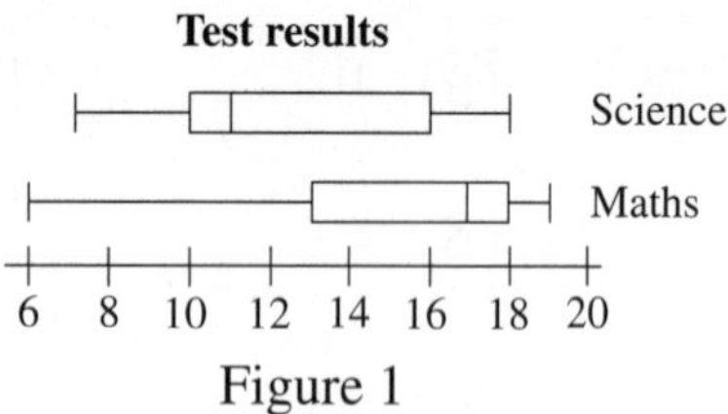

Figure 1

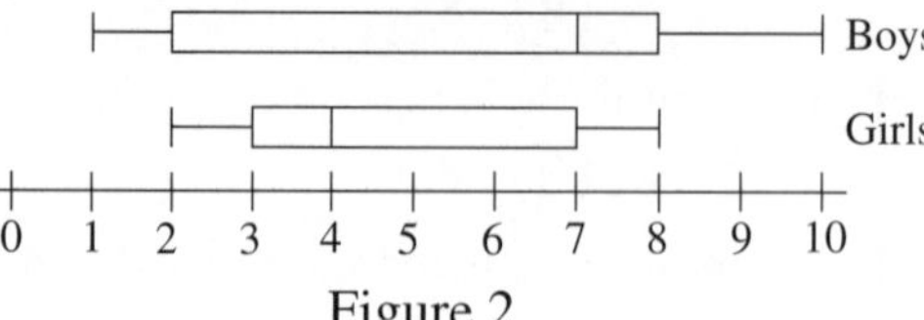

Figure 2

15 Using figure 1, comment, using calculations, on the differences in the:

a interquartile ranges **b** medians **c** skewness

6

16 Using figure 2, comment, using calculations, on the differences in the:

a interquartile ranges **b** medians **c** skewness

6

17 a The box plot records the following scores, listed in order: a, 6, b, 6, c, 8, d, 10.

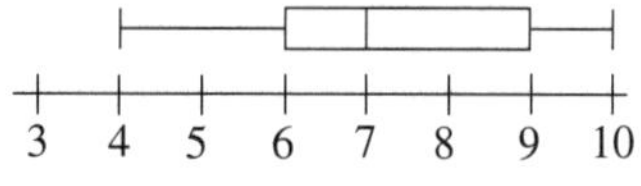

What is the value of each unknown pronumeral?

4

b The box plot records the following scores, listed in order: 8, 11, p, 11, q, 17, 18, r, 20, s.

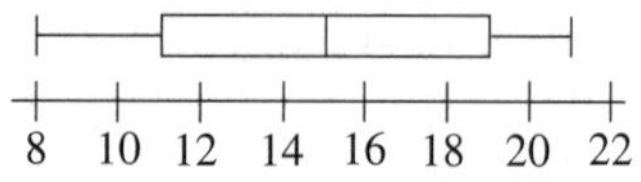

What is the value of each unknown pronumeral?

4

Total marks

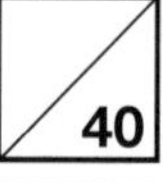

Probability

PRETEST

LEVEL 1

Time allowed: 10 minutes **Total marks: 10**

Marks

1 How many outcomes are there when a normal dice is rolled?

(A) 2 (B) 5 (C) 6 (D) 12 [1]

2 Which of these is the same as 25%?

(A) $\frac{1}{4}$ (B) $\frac{1}{25}$ (C) $\frac{2}{5}$ (D) $2\frac{1}{5}$ [1]

3 How many cards are in a normal deck of playing cards?

(A) 5 (B) 50 (C) 52 (D) 54 [1]

4 Ten balls numbered 1 to 10 are placed in a bag. How many of the balls have even numbers?

(A) 2 (B) 5 (C) 8 (D) 10 [1]

A dice is rolled 30 times and the results are listed in the table, which is used to answer questions 5 to 7.

Score	Number of times
1	4
2	7
3	6
4	
5	6
6	4

5 How many times was the number 4 rolled?

(A) 3 (B) 4 (C) 5 (D) 6 [1]

6 Which of these is the number of times an odd number was rolled?

(A) 3 (B) 12 (C) 15 (D) 16 [1]

7 Which of these is the number of times a number less than 3 was rolled?

(A) 2 (B) 6 (C) 11 (D) 17 [1]

8 Which of these is the value of $1 - \frac{1}{4}$?

(A) $\frac{1}{3}$ (B) $\frac{1}{2}$ (C) $\frac{2}{3}$ (D) $\frac{3}{4}$ [1]

9 Which of these is a red card in a normal pack of playing cards?

(A) king of clubs (B) seven of spades (C) ace of clubs (D) nine of diamonds 

10 Which of these is the number of consonants in the alphabet?

(A) 5 (B) 21 (C) 25 (D) 26 [1]

Total marks 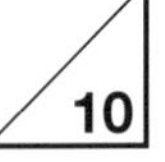

Probability

PRETEST — LEVELS 2 AND 3

Time allowed: 10 minutes — **Total marks: 10**

Marks

The numbers 1 to 15 are written on cards and the cards are used to answer questions 1 to 4.

1 Which of these is the number of cards with odd numbers?

(A) 2 (B) 7 (C) 8 (D) 9 — 1

2 Which of these is the number of cards with an even number less than 8?

(A) 3 (B) 4 (C) 6 (D) 7 — 1

3 Which of these is the number of cards with a prime number?

(A) 5 (B) 6 (C) 7 (D) 8 — 1

4 Which of these is the number of cards with a multiple of 4?

(A) 3 (B) 4 (C) 6 (D) 8 — 1

These statistical displays show the results of surveys and are used to answer questions 5 to 8.

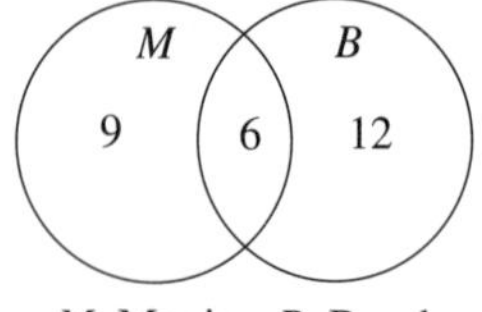

M: Movies B: Beach

Figure 1

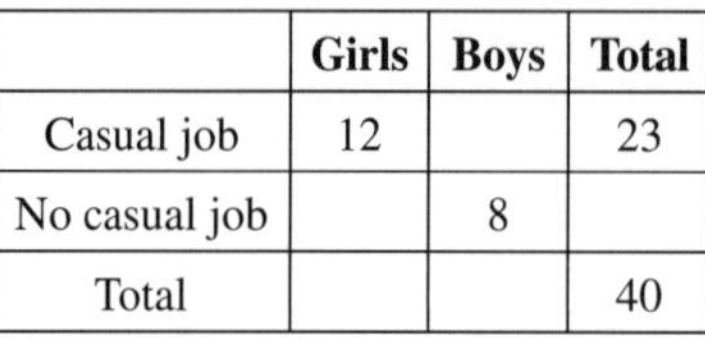

	Girls	Boys	Total
Casual job	12		23
No casual job		8	
Total			40

Figure 2

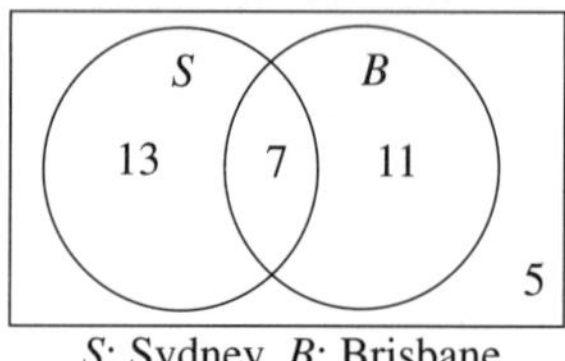

S: Sydney B: Brisbane

Figure 3

5 Using figure 1, what fraction of the people surveyed went to the movies?

(A) $\frac{1}{3}$ (B) $\frac{5}{9}$ (C) $\frac{2}{3}$ (D) $\frac{3}{4}$ — 1

6 Using figure 2, which of these is the total number of girls surveyed?

(A) 12 (B) 18 (C) 20 (D) 21 — 1

7 Using figure 2, what percentage of the total surveyed were boys with a casual job?

(A) 11% (B) 20% (C) 27.5% (D) 57.9% — 1

8 Using figure 3, how many people surveyed had **not** been to Sydney?

(A) 5 (B) 16 (C) 23 (D) 36 — 1

9 The probability of choosing a red ball from a bag containing red, blue and green balls is $\frac{7}{11}$. Which of these is the probability of the complementary event?

(A) $\frac{1}{11}$ (B) $\frac{1}{7}$ (C) $\frac{4}{11}$ (D) $\frac{2}{3}$ — 1

10 The probability of a soccer team winning a certain game is 0.38. If the probability of not losing is 0.55, which of these is the probability of the team drawing the game?

(A) 0.17 (B) 0.33 (C) 0.5 (D) 0.93 — 1

Total marks /10

Probability

LEVEL 1 TEST — PART A

Time allowed: 40 minutes

Marks

A bag contains 12 red cards, five green cards and three blue cards. This information is used to answer questions 1 to 3.

1 Which of these is the probability of choosing a green card?

(A) 0.05 (B) 0.25 (C) 0.33 (D) 0.5 — 1

2 Which of these is the probability of **not** choosing a blue card?

(A) 0.17 (B) 0.66 (C) 0.7 (D) 0.85 — 1

3 If five more blue cards are added to the bag, what is the probability of choosing a red card?

(A) 0.2 (B) 0.33 (C) 0.48 (D) 0.55 — 1

A box contains 20 balls numbered 1 to 20. This information is used to answer questions 4 to 6.

4 What is the probability of selecting a number less than 10?

(A) 0.09 (B) 0.2 (C) 0.45 (D) 0.5 — 1

5 Which of these is the probability of selecting an odd number greater than 12?

(A) 0.2 (B) 0.4 (C) 0.42 (D) 0.6 — 1

6 What is the probability of selecting a ball that is a factor of 24?

(A) 0.29 (B) 0.3 (C) 0.35 (D) 0.7 — 1

In a raffle, tickets numbered 1 to 100 are sold. This information is used to answer questions 7 to 9.

7 How many tickets are between 63 and 68?

(A) 4 (B) 5 (C) 6 (D) 7 — 1

8 What is the probability that the winning ticket is even and higher than 80?

(A) 0.1 (B) 0.18 (C) 0.2 (D) 0.81 — 1

9 Which of these is the probability that the winning ticket is **not** less than 10?

(A) 0.09 (B) 0.1 (C) 0.9 (D) 0.91 — 1

10 A dice is rolled 48 times. What is the likely number of times a 4 is rolled?

(A) 4 (B) 6 (C) 8 (D) 12 — 1

11 The relative frequency table is shown for a set of scores.

Score	Frequency	Relative *F*
11	5	0.25
12	9	0.45
13	6	a
Total	20	1.00

Which of these is the value of a?

(A) 0.03 (B) 0.06 (C) 0.3 (D) 0.6 — 1

A sticker is placed on a die, changing the 6 to a 5. Use this information to answer questions 12 and 13.

12 Which of these is the probability of rolling a 5?

(A) $\frac{1}{3}$ (B) $\frac{2}{5}$ (C) $\frac{1}{2}$ (D) $\frac{2}{3}$ — 1

13 Which of these is the probability of rolling an even number?

(A) $\frac{1}{3}$ (B) $\frac{2}{5}$ (C) $\frac{1}{2}$ (D) $\frac{2}{3}$ — 1

LEVEL 1 TEST — PART B

Write answers and working in the space provided.

Marks

14 A card is selected from a normal pack of playing cards. What is the probability of selecting:

a a red ace? ________ **b** a nine? ________ **c** a spade? ________ 3

d a 5 or 6? ________ **e** a picture card? ________ **f** **not** a diamond? ________ 3

15 Ten coloured discs are placed in a bag. There are three green discs, five yellow discs and the rest are purple. If a disc is selected from the bag at random, what is the probability that it is:

a purple? ________ **b** green or yellow? ________ **c** **not** green? ________ 3

16 A group of students rolled a pair of dice a total of 100 times and recorded the sum of their uppermost faces in the table.

Total	Frequency	Relative frequency
2	3	
3	5	
4	8	
5	11	
6	14	
7	a	
8	13	
9	10	
10	7	
11	6	
12	4	

a What is the value of a? ________ 1

b Complete the table. 1

c What is the probability of rolling a total:

i of 10? ________ 1

ii less than 6? ________ 1

iii greater than 10? ________ 1

iv at least 8? ________ 1

Three student surveys were conducted and the results displayed below.

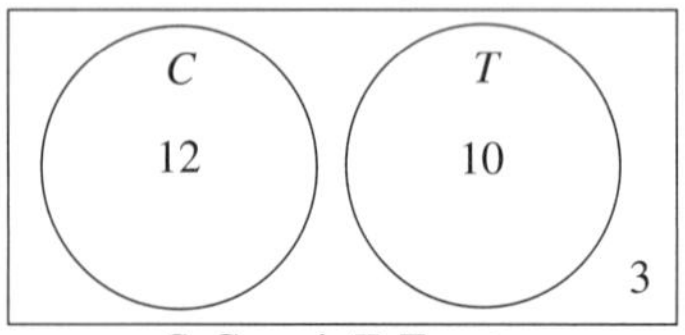

C: Cereal *T*: Toast

Figure 1

	Girls	Boys	Total
Left-handed	5	7	12
Right-handed	12	16	28
Total	17	23	40

Figure 2

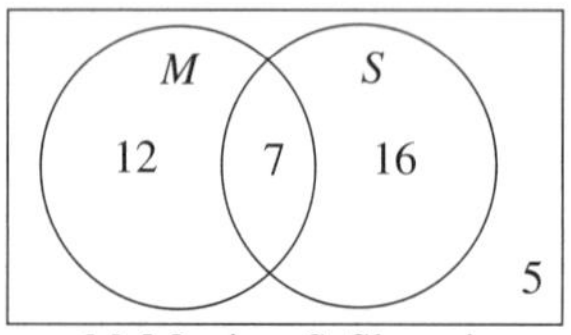

M: Movies *S*: Shopping

Figure 3

17 Using figure 1, what is the probability that a student chosen at random:

a had cereal? ________ **b** did **not** have cereal or toast? ________ 2

c had cereal and toast? ________ **d** had cereal or toast? ________ 2

18 Using figure 2, what is the probability that a student chosen at random:

a is a girl? ________ **b** is a right-handed boy? ________ 2

c is left-handed? ________ **d** is **not** a right-handed girl? ________ 2

19 Using figure 3, what is the probability that a student:

a saw a movie and shopped? ________ **b** shopped but did **not** see a movie? ________ 2

c did **not** see a movie? ________ **d** neither shopped nor saw a movie? ________ 2

Total marks /40

Probability

LEVEL 2 TEST PART A

Time allowed: 40 minutes

Marks

1 The faces on a 10-sided dice are numbered 1 to 10. What is the probability of **not** rolling a 5?

(A) 0 (B) 0.1 (C) 0.5 (D) 0.9 1

A game has two outcomes, win or lose, and the probability of winning is 0.3. This information is used to answer questions 2 to 4.

2 What is the probability of winning two consecutive games?

(A) 0.06 (B) 0.09 (C) 0.5 (D) 0.6 1

3 What is the probability of losing two consecutive games?

(A) 0.09 (B) 0.14 (C) 0.33 (D) 0.49 1

4 What is the probability of winning at least one game out of two played?

(A) 0.39 (B) 0.42 (C) 0.49 (D) 0.51 1

The figures are used to answer questions 5 to 13.

Vehicles	Manual	Automatic
Sedan	6	18
SUV	12	24

Figure 1

School uniform	Yr 8	Yr 9	Yr 10
Like	40	50	60
Dislike	40	30	20

Figure 2

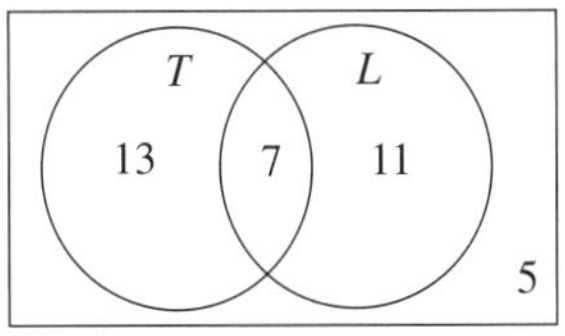

Figure 3

5 Using figure 1, which of these is the probability of choosing an automatic sedan?

(A) 0.18 (B) 0.25 (C) 0.3 (D) 0.5625 1

6 Using figure 1, which of these is the probability of choosing an SUV?

(A) 0.25 (B) 0.36 (C) 0.5 (D) 0.6 1

7 Using figure 1, which of these is the probability of **not** choosing a manual SUV?

(A) 0.4 (B) 0.75 (C) 0.8 (D) 0.88 1

8 Using figure 1, if a sedan is chosen, what is the probability that it is an automatic?

(A) 0.18 (B) 0.33 (C) 0.5 (D) 0.75 1

9 Using figure 2, what is the probability of choosing a Year 9 student who likes the uniform?

(A) $\frac{5}{24}$ (B) $\frac{1}{6}$ (C) $\frac{1}{2}$ (D) $\frac{5}{8}$ 1

10 Using figure 2, if a Year 8 student is chosen, what is the probability that they do **not** like the uniform?

(A) $\frac{1}{40}$ (B) $\frac{1}{6}$ (C) $\frac{1}{2}$ (D) $\frac{4}{9}$ 1

11 Using figure 2, if a student who likes the uniform is chosen, what is the probability they are **not** in Year 8?

(A) $\frac{2}{11}$ (B) $\frac{11}{24}$ (C) $\frac{2}{3}$ (D) $\frac{11}{15}$ 1

12 Using figure 3, which of these is the probability of choosing a person who has a laptop and also has a tablet?

(A) $\frac{7}{36}$ (B) $\frac{7}{31}$ (C) $\frac{11}{36}$ (D) $\frac{7}{11}$ 1

13 Using figure 3, if a person without a tablet is chosen, what is the probability that they do **not** have a laptop?

(A) $\frac{5}{18}$ (B) $\frac{5}{16}$ (C) $\frac{1}{2}$ (D) $\frac{11}{18}$ 1

Probability

LEVEL 2 TEST — PART B

Write answers and working in the space provided.

Marks

The tree diagrams are used to answer questions 14, 15 and 16.

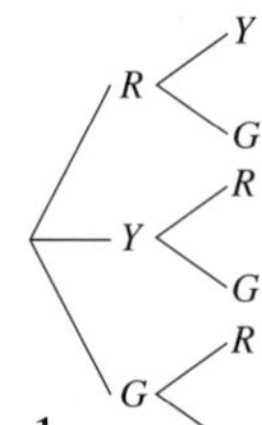

Figure 1

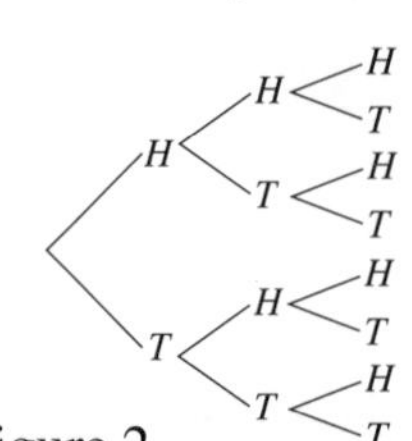

Figure 2

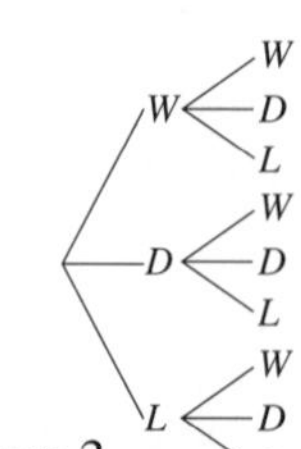

Figure 3

14 Using figure 1, two balls are chosen from a bag containing a red ball, a yellow ball and a green ball. What is the probability of choosing a:

a red and a green ball? ______ **b** yellow ball? ______ **c** red ball or a yellow ball? ______ 3

15 Using figure 2, three coins are tossed. What is the probability of tossing:

a three heads? ______ **b** three heads or three tails? ______ **c** at least one head? ______ 3

d no tails? ______ **e** at least two heads? ______ **f** at most two tails? ______ 3

16 Using figure 3, the results (*W*: win, *D*: draw, *L*: loss) of two games played by a soccer team are recorded. Assuming each result is equally likely, what is the probability of:

a two wins? ______ **b** a win followed by a loss? ______ **c** a draw and a loss? ______ 3

d at least a win? ______ **e** no wins? ______ **f** no wins or draws? ______ 3

Two dice are rolled and the results shown in the tables which are used to answer questions 17 and 18.

Figure 4

+	1	2	3	4	5	6
1	2	3	4	5	6	7
2	3	4	5	6	7	8
3	4	5	6	7	8	9
4	5	6	7	8	9	10
5	6	7	8	9	10	11
6	7	8	9	10	11	12

Figure 5

×	1	2	3	4	5	6
1	1	2	3	4	5	6
2	2	4	6	8	10	12
3	3	6	9	12	15	18
4	4	8	12	16	20	24
5	5	10	15	20	25	30
6	6	12	18	24	30	36

17 Using figure 4, the results on the dice are added. What is the probability of a result:

a of 4? ______ **b** of 12? ______ **c** less than 5? ______ 3

d greater than 9? ______ **e** which is even? ______ **f** which is not 7? ______ 3

18 Using figure 5, the results on the dice are multiplied. What is the probability of a result:

a of 36? ______ **b** which is odd? ______ **c** which is prime? ______ 3

d less than 12? ______ **e** which is a multiple of 3? ______ **f** which is a factor of 36? ______ 3

Total marks /40

Probability

LEVEL 3 TEST — PART A

Time allowed: 40 minutes

Marks

1 A bag contains 20 marbles. The probability of selecting a green marble is 0.4. How many green marbles are in the bag?

(A) 4 (B) 6 (C) 8 (D) 9 — 1

In an experiment, a coin is tossed and a dice is rolled. This information is used to answer questions 2 and 3.

2 What is the probability of a head and a 5?

(A) $\frac{1}{12}$ (B) $\frac{1}{10}$ (C) $\frac{1}{2}$ (D) $\frac{2}{3}$ — 1

3 What is the probability of a tail and an even number?

(A) 0.25 (B) 0.5 (C) 0.6 (D) 1 — 1

Two balls are chosen from a bag containing three red balls, two green balls and a white ball. This information is used to answer questions 4 to 6.

4 What is the probability of two red balls?

(A) $\frac{1}{15}$ (B) $\frac{1}{5}$ (C) $\frac{1}{4}$ (D) $\frac{1}{2}$ — 1

5 What is the probability of two balls of the same colour?

(A) $\frac{2}{15}$ (B) $\frac{4}{15}$ (C) $\frac{8}{15}$ (D) $\frac{2}{3}$ — 1

6 What is the probability of a green ball and a red ball?

(A) $\frac{1}{5}$ (B) $\frac{1}{3}$ (C) $\frac{4}{15}$ (D) $\frac{2}{5}$ — 1

Twenty tickets are sold in a raffle where there are three prizes. Use this information to answer questions 7 to 9.

7 Todd buys two tickets. What is the probability he wins first and second prizes?

(A) $\frac{1}{380}$ (B) $\frac{1}{200}$ (C) $\frac{1}{190}$ (D) $\frac{1}{10}$ — 1

8 Elizabeth buys one ticket. What is the probability she wins third prize?

(A) $\frac{1}{20}$ (B) $\frac{1}{18}$ (C) $\frac{1}{17}$ (D) $\frac{1}{3}$ — 1

9 Donald buys half the tickets sold. What is the probability he does **not** win a prize?

(A) $\frac{36}{323}$ (B) $\frac{3}{17}$ (C) $\frac{2}{19}$ (D) $\frac{17}{20}$ — 1

Two dice are rolled and the difference in their numbers is used to answer questions 10 and 11.

10 What is the probability that the difference is zero?

(A) $\frac{1}{36}$ (B) $\frac{1}{12}$ (C) $\frac{5}{36}$ (D) $\frac{1}{6}$ — 1

11 What is the probability that the difference is five?

(A) $\frac{1}{36}$ (B) $\frac{1}{18}$ (C) $\frac{5}{36}$ (D) $\frac{1}{6}$ — 1

A coin is tossed four times and the results are used to answer questions 12 and 13.

12 Which of these is the number of outcomes?

(A) 8 (B) 12 (C) 16 (D) 32 — 1

13 What is the probability of four heads or four tails?

(A) $\frac{1}{16}$ (B) $\frac{1}{8}$ (C) $\frac{1}{4}$ (D) $\frac{1}{2}$ — 1

LEVEL 3 TEST

PART B

Write answers and working in the space provided.

Marks

The diagrams will be used to answer questions 14 to 17.

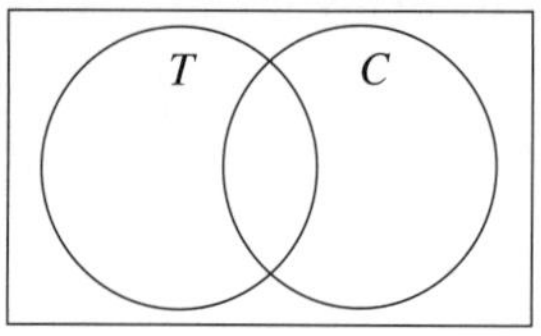

T: Tea
C: Coffee

Figure 1

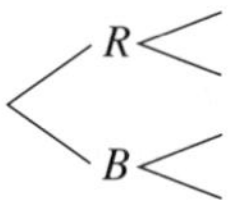

Figure 2

Driver's licence	Under 20	20 and over
Yes		
No		

Figure 3

14 A survey of 40 people was conducted to find the number who regularly drink tea and coffee. It was found six drink neither, 18 drink tea and 26 drink coffee.

a Use the information to complete the Venn diagram in figure 1. 1

b What is the probability of a person drinking both tea and coffee? ______ 1

c What is the probability of a person **not** drinking coffee? ______ 1

15 A bag contains four red balls and two blue balls. A ball is chosen at random and then is replaced by a ball of the opposite colour.

a Use the information to complete the tree diagram in figure 2. 1

b What is the probability of two different coloured balls being chosen? ______ 1

c What is the probability of a blue ball given that a red ball has been chosen? ______ 1

16 A survey of 60 people was conducted to find the number of people who held a driver's licence. It was found that 40% of the under-20-year-olds had a licence. Also, one-third of those surveyed did not have a licence and two-thirds were aged 20 years and over.

a Use the information to complete the table in figure 3. 1

b What is the probability of randomly selecting a licensed under-20-year-old? ______ 1

c What is the probability that a licensed driver is under 20 years old? ______ 1

17 A bag contains nine balls numbered 1 to 9. Two balls are chosen at random. What is the probability that:

a both balls are even? ______

b both balls are less than 4? ______

c the sum of the balls is 17? ______

6

d exactly one ball is 7? ______

e at least one ball is prime? ______

f the first ball is even and the second ball is odd? ______

6

18 Bob, Bill and Ben are racing each other. Bob is twice more likely to win than either of the other two boys. What is the probability that Bill wins the race? ______ 2

19 Two dice are rolled. What is the probability that a 6 only appears once? ______ 2

20 At the same time three coins are tossed and the result recorded. If this experiment occurs a total of 120 times, on how many occasions is it expected that three heads will appear? ______ 2

Total marks /40

Statistics and Probability Test: Level 1

STRAND TEST — PART A

Time allowed: 50 minutes

Marks

The displays are used to answer questions 1 to 7.

Class	Class centre	Frequency
13–19	16	3
20–26	*p*	5
27–33	30	8
34–40	37	*q*
Total		20

Figure 1

Stem	Leaf
3	*r*48
4	11*s*47
5	258

Figure 2

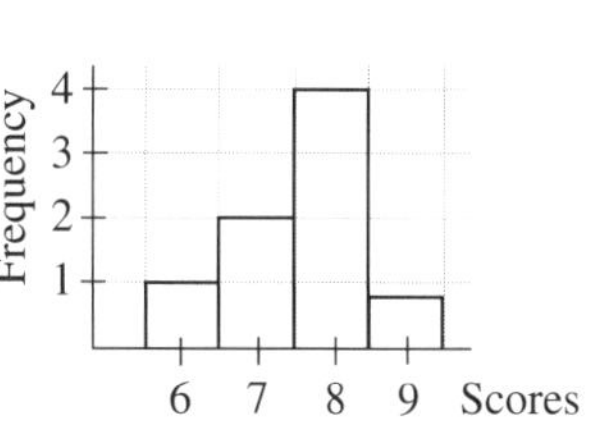

Figure 3

1 Using figure 1, which of these is the value of *p*?

Ⓐ 21 Ⓑ 22 Ⓒ 23 Ⓓ 25 [1]

2 Using figure 1, which of these is the value of *q*?

Ⓐ 1 Ⓑ 2 Ⓒ 3 Ⓓ 4 [1]

3 Using figure 1, which of these is the modal class?

Ⓐ 13–19 Ⓑ 20–26 Ⓒ 27–33 Ⓓ 34–40 [1]

4 Using figure 2, if the range is 26, which of these is the value of *r*?

Ⓐ 1 Ⓑ 2 Ⓒ 3 Ⓓ 4 [1]

5 Using figure 2, if the median is 42, which of these is the value of *s*?

Ⓐ 1 Ⓑ 2 Ⓒ 3 Ⓓ 4 [1]

6 Using figure 3, which of these is the range?

Ⓐ 1 Ⓑ 2 Ⓒ 3 Ⓓ 4 [1]

7 Using figure 3, which of these is the total number of scores?

Ⓐ 4 Ⓑ 5 Ⓒ 6 Ⓓ 8 [1]

A box contains 20 coloured balls. Seven of the balls are red, six are green, five are blue and the remainder are yellow. This information is used to answer questions 8 to 13.

8 What is the probability of selecting a blue ball?

Ⓐ 0.2 Ⓑ 0.25 Ⓒ 0.4 Ⓓ 0.5 [1]

9 Which of these is the probability of selecting a yellow ball?

Ⓐ 0.1 Ⓑ 0.2 Ⓒ 0.25 Ⓓ 0.4 [1]

10 What is the probability of selecting a red ball or a green ball?

Ⓐ 0.25 Ⓑ 0.35 Ⓒ 0.5 Ⓓ 0.65 [1]

11 Which of these is the probability of **not** selecting a yellow ball?

Ⓐ 0.6 Ⓑ 0.7 Ⓒ 0.8 Ⓓ 0.9 [1]

A dice is rolled and the result used to answer questions 12 and 13.

12 Which of these is the probability of rolling an odd number less than 4?

Ⓐ $\frac{1}{3}$ Ⓑ $\frac{1}{2}$ Ⓒ $\frac{2}{3}$ Ⓓ $\frac{5}{6}$ [1]

13 An even number is rolled. What is the probability that it is 2?

Ⓐ $\frac{1}{6}$ Ⓑ $\frac{1}{3}$ Ⓒ $\frac{1}{2}$ Ⓓ $\frac{2}{3}$ [1]

Statistics and Probability Test: Level 1

STRAND TEST — PART B

Write answers and working in the space provided.

Marks

14 The stem-and-leaf display shows the results of two tests.

Test results

Science		English
631	1	479
866310	2	02455
0	3	38

a What is the mode of the:

i Science results? ____________ **ii** English results? ____________ 2

b What is the range of the:

i Science results? ____________ **ii** English results? ____________ 2

c What is the median of the:

i Science results? ____________ **ii** English results? ____________ 2

15 Here is a list of waiting times (in minutes) for a medical surgery:

15, 20, 17, 8, 10, 17, 23, 22, 17, 5, 12, 18, 20, 23, 17.

Complete the table.

Class	Class	Frequency
5–9		
10–14		
15–19		
20–24		
Total		

2

16 This is a list of the number of goals scored in 10 games of soccer in a season:

5, 0, 3, 4, 7, 2, 5, 3, 6, 5.

Find the:

a range ____________ **b** mode ____________ **c** mean ____________ 3

d median ____________ **e** lower quartile ____________ **f** upper quartile ____________ 3

17

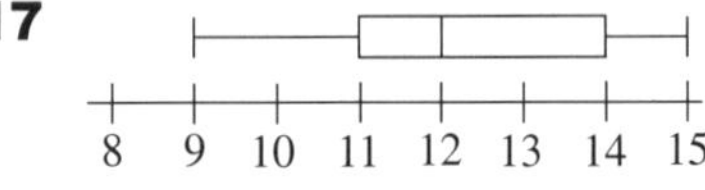

What is the:

a median? ____________ **b** interquartile range? ____________ 2

18 The results of an analysis of the vowels used in a paragraph of writing are listed in the relative frequency table below.

Vowels	Frequency	Relative frequency
a	26	
e	32	
i	18	
o		
u	8	
Total	100	

a Complete the table. 3

b According to the table, if a vowel is chosen at random, what is the probability of:

i selecting an *a*? ____________ **ii** selecting an *a* or an *e*? ____________ 2

iii **not** selecting an *o*? ____________ 1

Statistics and Probability Test: Level 1

STRAND TEST — PART B

Write answers and working in the space provided.

Marks

19 A survey found the number of people who check their Facebook and Instagram accounts daily.

F: Facebook
I: Instagram

a How many people were surveyed? ________________ [1]

b If a person is chosen at random, what is the probability that they:

i check Facebook? ________________

ii check Facebook but **not** Instagram? ________________ [2]

iii check neither? ________________

iv do **not** check Facebook? ________________ [2]

20 The travel experience of students in a class was surveyed and the results listed in the table below.

	Boys	Girls	Total
Overseas	8	12	
Not overseas	6	4	
Total			

a Complete the table. [2]

b If a student is chosen at random, what is the probability that they:

i are a boy who has travelled overseas? ________________

ii are a girl who has **not** travelled overseas? ________________ [2]

iii have travelled overseas? ________________

iv are a girl? ________________ [2]

21 A card is chosen at random from a pack of normal playing cards.

What is the probability of selecting:

a a heart? ________________

b a red seven? ________________ [2]

c a king or a queen? ________________

d **not** an ace or a picture card? ________________ [2]

Total marks

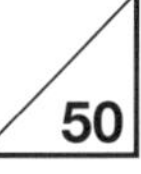

Statistics and Probability Test: Level 2

STRAND TEST — PART A

Time allowed: 50 minutes

Marks

The scores 3, 5, 7, 9, 11, 13 are used to answer questions 1 to 3.

1 Which of these is the interquartile range of the scores?

(A) 3 (B) 4 (C) 5 (D) 6 — 1

2 If each of the scores is increased by 2, what is the new range?

(A) 10 (B) 12 (C) 13 (D) 15 — 1

3 If the minimum score is ignored, what is the new median?

(A) 7 (B) 8 (C) 9 (D) 10 — 1

4 If another score is included, the mean decreases by 1. Which of these is the new score?

(A) 1 (B) 3 (C) 6 (D) 7 — 1

The displays are used to answer questions 5 to 13.

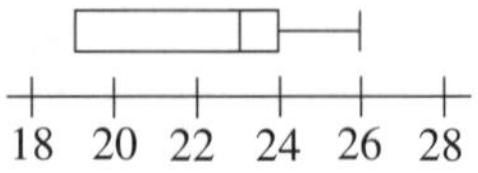

Figure 1

Swim 200 m	Yr 8	Yr 9	Yr 10
Yes	14	12	5
No	6	6	7

Figure 2

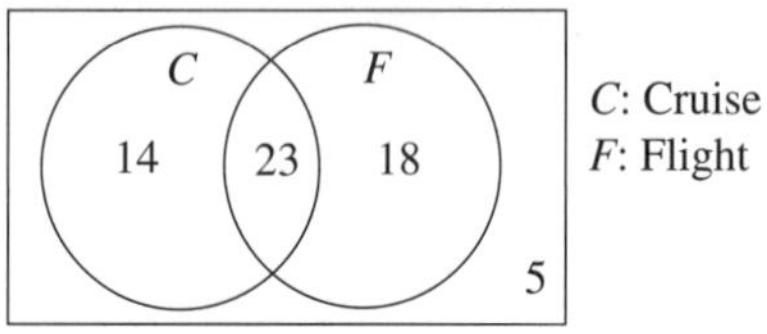

Figure 3

5 Using figure 1, which percentage of the scores is between 23 and 26?

(A) 3% (B) 25% (C) 50% (D) 75% — 1

6 Using figure 1, which of these is the minimum score?

(A) 19 (B) 19.5 (C) 23 (D) 28 — 1

7 Using figure 2, how many students surveyed were capable of swimming 200 m?

(A) 31 (B) 33 (C) 45 (D) 50 — 1

8 Using figure 2, what is the probability of choosing a Year 9 student who can swim 200 m?

(A) 0.12 (B) 0.24 (C) 0.36 (D) 0.62 — 1

9 Using figure 2, if a Year 8 student is chosen, what is the probability that they can swim 200 m?

(A) 0.14 (B) 0.28 (C) 0.56 (D) 0.7 — 1

10 Using figure 2, what is the probability that a randomly chosen student is not in Year 9?

(A) 0.12 (B) 0.64 (C) 0.82 (D) 0.88 — 1

11 Using figure 3, what is the probability that a person has taken a cruise?

(A) $\frac{7}{30}$ (B) $\frac{23}{60}$ (C) $\frac{37}{60}$ (D) $\frac{11}{12}$ — 1

12 Using figure 3, what is the probability that a person has **not** taken a flight?

(A) $\frac{7}{30}$ (B) $\frac{19}{60}$ (C) $\frac{7}{10}$ (D) $\frac{41}{60}$ — 1

13 Using figure 3, what is the probability that a person who has taken a flight has **not** taken a cruise?

(A) $\frac{18}{41}$ (B) $\frac{3}{10}$ (C) $\frac{23}{41}$ (D) $\frac{28}{41}$ — 1

Statistics and Probability Test: Level 2

STRAND TEST

PART B

Write answers and working in the space provided.

Marks

14

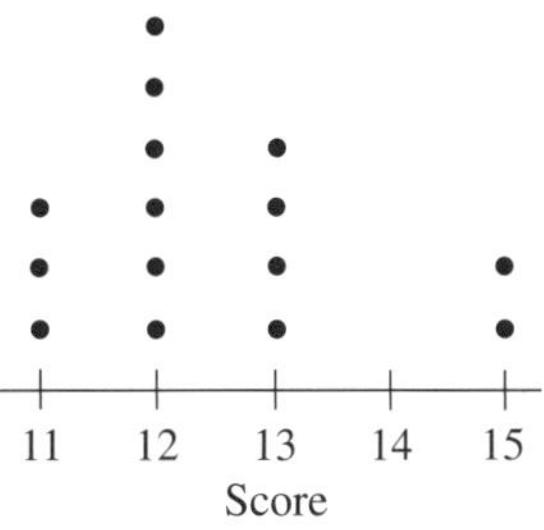

Find the:

a mode ______ **b** median ______ **c** lower quartile ______ 3

d upper quartile ______ **e** interquartile range ______ **f** mean, to two decimal places ______ 3

15 The diagram shows a cumulative frequency histogram and polygon.

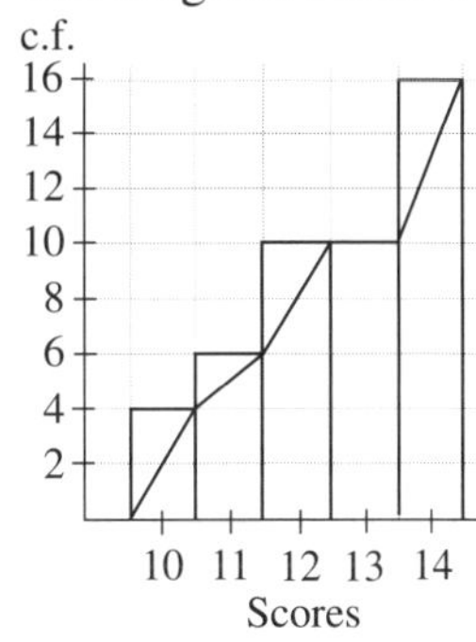

Score (x)	Frequency (f)	*fx*
10		
11		
12		
13		
14		
Total		

a Complete the frequency table. 2

b Find the:

i mode ______ **ii** mean ______ **iii** median ______ 3

16 Complete 5-point summaries and draw box plots for the following:

a 11, 15, 13, 16, 14, 12, 14

10 11 12 13 14 15 16 17

______
______
______
______ 3

b 28, 22, 18, 30, 26, 24, 28, 31, 32, 20

18 20 22 24 26 28 30 32

______
______
______
______ 3

Statistics and Probability Test: Level 2

STRAND TEST PART B

Write answers and working in the space provided.

Marks

17 The tree diagram is used to list the sample space of a family of three children.

G

B

a Complete the tree diagram. 2

b What is the probability of:

i three children of the same gender?

ii two boys and a girl? 2

iii at least two girls?

iv a majority of boys? 2

18 A coin is tossed and a dice is rolled.

a How many outcomes are in the sample space? 1

b What is the probability of a:

i head and a 3?

ii tail and a number smaller than 5?

iii head and a multiple of 3? 3

19 A bag contains three red balls and three blue balls. A ball is chosen at random, its colour noted and it is placed back into the bag before a second ball is chosen.

What is the probability of:

a choosing two red balls?

b different colours? 4

20 Two dice are rolled and the higher of the two numbers recorded as the score.

	1	2	3	4	5	6
1						
2						
3						
4						
5						
6						

a Complete the table for the possible scores. 2

b What is the probability that the score is:

i even?

ii less than 4? 2

iii prime?

iv greater than 3 and even? 2

Total marks /50

Statistics and Probability Test: Level 3

STRAND TEST — PART A

Time allowed: 50 minutes

Marks

The diagrams are used to answer questions 1 to 13.

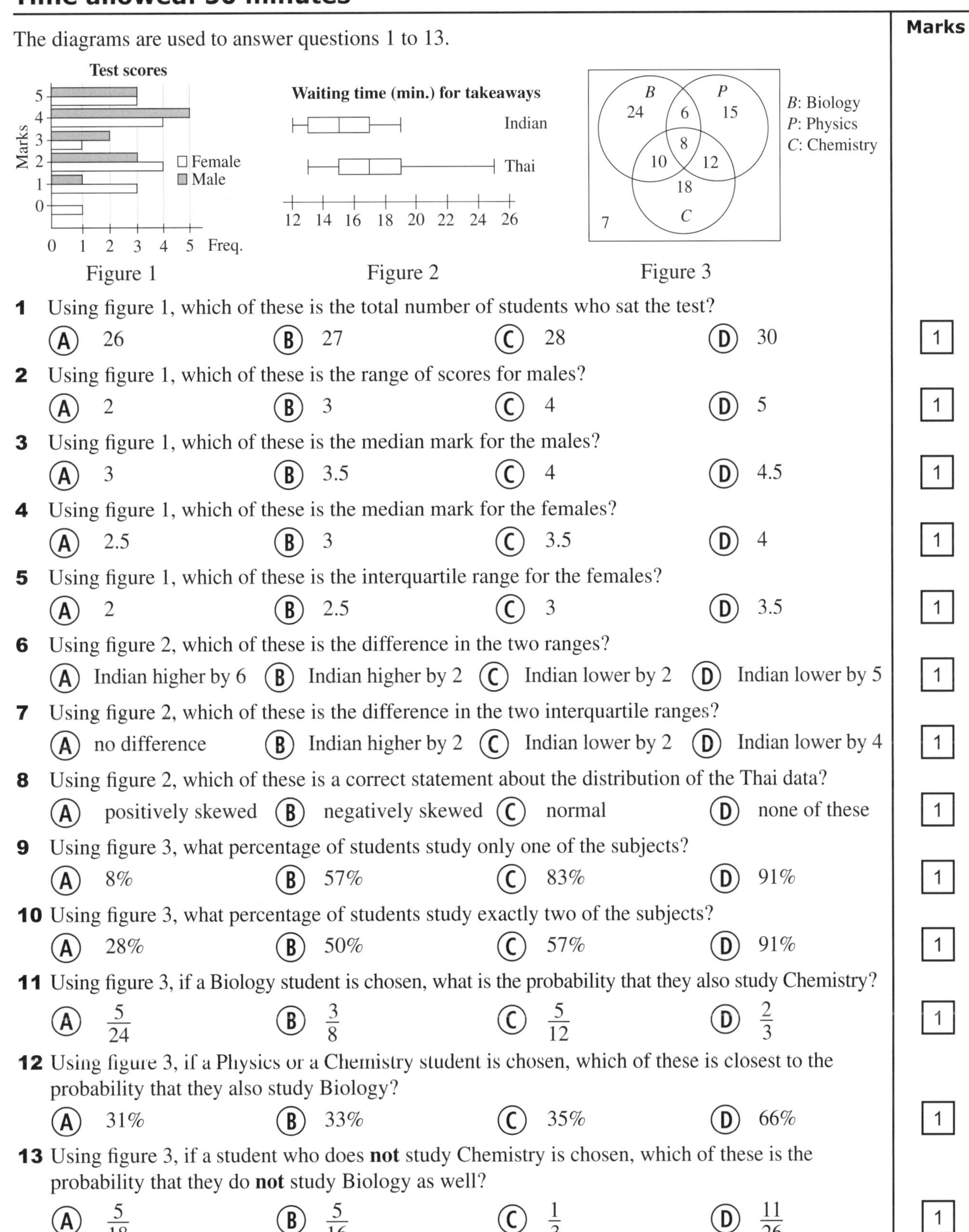

Figure 1 — Figure 2 — Figure 3

1 Using figure 1, which of these is the total number of students who sat the test?
(A) 26 (B) 27 (C) 28 (D) 30 — 1

2 Using figure 1, which of these is the range of scores for males?
(A) 2 (B) 3 (C) 4 (D) 5 — 1

3 Using figure 1, which of these is the median mark for the males?
(A) 3 (B) 3.5 (C) 4 (D) 4.5 — 1

4 Using figure 1, which of these is the median mark for the females?
(A) 2.5 (B) 3 (C) 3.5 (D) 4 — 1

5 Using figure 1, which of these is the interquartile range for the females?
(A) 2 (B) 2.5 (C) 3 (D) 3.5 — 1

6 Using figure 2, which of these is the difference in the two ranges?
(A) Indian higher by 6 (B) Indian higher by 2 (C) Indian lower by 2 (D) Indian lower by 5 — 1

7 Using figure 2, which of these is the difference in the two interquartile ranges?
(A) no difference (B) Indian higher by 2 (C) Indian lower by 2 (D) Indian lower by 4 — 1

8 Using figure 2, which of these is a correct statement about the distribution of the Thai data?
(A) positively skewed (B) negatively skewed (C) normal (D) none of these — 1

9 Using figure 3, what percentage of students study only one of the subjects?
(A) 8% (B) 57% (C) 83% (D) 91% — 1

10 Using figure 3, what percentage of students study exactly two of the subjects?
(A) 28% (B) 50% (C) 57% (D) 91% — 1

11 Using figure 3, if a Biology student is chosen, what is the probability that they also study Chemistry?
(A) $\frac{5}{24}$ (B) $\frac{3}{8}$ (C) $\frac{5}{12}$ (D) $\frac{2}{3}$ — 1

12 Using figure 3, if a Physics or a Chemistry student is chosen, which of these is closest to the probability that they also study Biology?
(A) 31% (B) 33% (C) 35% (D) 66% — 1

13 Using figure 3, if a student who does **not** study Chemistry is chosen, which of these is the probability that they do **not** study Biology as well?
(A) $\frac{5}{18}$ (B) $\frac{5}{16}$ (C) $\frac{1}{3}$ (D) $\frac{11}{26}$ — 1

Statistics and Probability Test: Level 3

STRAND TEST — PART B

Write answers and working in the space provided.

	Marks
14 The number of home deliveries made by two different suburban supermarkets in a 10-day period is recorded below. *A*: 11, 16, 23, 17, 15, 31, 40, 38, 27, 42 *B*: 23, 26, 45, 37, 35, 27, 21, 14, 38, 35 **a** Display the data in a back-to-back stem-and-leaf display.	3
b Use the display to comment on the differences between the number of deliveries from the two supermarkets. **i** range **ii** median **iii** interquartile range	6
15 The times run for a 100-m race by Year 9 students are recorded below. Girls: 13.4, 13.5, 14.1, 14.6, 14.0, 12.6, 14.4 Boys: 12.2, 13.6, 13.4, 12.8, 12.6, 12.7, 14.2 **a** Complete a 5-point summary for each list. **i** Girls **ii** Boys	6
b Use the scale below to record the box plots for the data. Girls Boys 12.2 12.6 13.0 13.4 13.8 14.2 14.6	2
c Comment on the differences in the: **i** interquartile range **ii** skewness	4
16 The box plot records the following scores, listed in order: *a*, 19, 22, *b*, 25, 25, *c*, 28, 28, *d*, 29, 30. 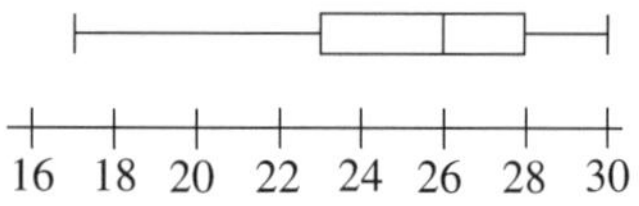 16 18 20 22 24 26 28 30 What is the value of each unknown pronumeral?	4

Statistics and Probability Test: Level 3

STRAND TEST — PART B

Write answers and working in the space provided.

Marks

17 Thirty tickets in a raffle are sold. There are three prizes to be drawn.

a Daniel buys one ticket. What is the probability that he wins:

i first prize? **ii** second prize? **iii** third prize?

3

b Adam buys two tickets. What is the probability that he:

i does **not** win a prize? **ii** wins exactly one prize?

2

18 On a normal dice the number 6 is replaced with a 7. If the dice is rolled twice, what is the probability that the total is:

a 12? **b** greater than 9?

2

19 A box contains 60 balls. The balls are either red, blue or green.

a Suppose the probability of choosing a red ball is 0.4. How many balls are **not** red?

1

b Suppose there are twice as many blue balls as green balls and the probability of **not** selecting a red ball is 0.7. How many blue balls are in the bag?

2

20 Ten balls numbered from 1 to 10 are placed in a bag. A ball is chosen at random and is discarded before a second ball is chosen.

a What is the probability that the first two balls are even?

1

b If this action continues, what is the probability that all the even-numbered balls will be selected before an odd ball is chosen?

1

Total marks /50

Exam Paper: Level 1

Instructions for all parts • Attempt all questions.
Time allowed: $1\frac{1}{2}$ hours • Calculators are allowed.

Part A: Allow about 20 minutes for this part.
Part B: Allow about 30 minutes for this part.
Part C: Allow about 40 minutes for this part.

Total marks: 100

EXAM PAPER: LEVEL 1 — PART A

Fill in only one circle for each question.

Marks

1 Which of these is $\frac{11.739}{4.8931}$, correct to two decimal places? [1]

(A) 2.39 (B) 2.40 (C) 2.50 (D) 2.59

2 Which of these is the largest? [1]

(A) $\frac{5}{9}$ (B) $0.5\dot{4}$ (C) 55% (D) 0.506

3 Elizabeth is paid at an hourly rate of $21.30. How much will she be paid if she works for 6 hours at time-and-a-half? [1]

(A) $159.75 (B) $191.70 (C) $255.60 (D) $319.50

4 What is the simple interest on $3600 invested for 5 years at 3% p.a. interest? [1]

(A) $504 (B) $540 (C) $573 (D) $4140

5 Simplify $5p - p + 2p$? [1]

(A) $2p$ (B) $4p$ (C) $6p$ (D) $8p$

6 Simplify $\frac{m}{4} \times \frac{m}{2}$. [1]

(A) $\frac{1}{8}$ (B) $\frac{m^2}{2}$ (C) $\frac{m^2}{6}$ (D) $\frac{m^2}{8}$

7 Simplify $(a^4)^2$. [1]

(A) a^6 (B) a^8 (C) $6a$ (D) $8a$

8 Which of these is the value of 4^5? [1]

(A) 20 (B) 625 (C) 1024 (D) 44 444

9 Solve the equation $\frac{a+2}{4} = 6$. [1]

(A) $a = 8$ (B) $a = 12$ (C) $a = 22$ (D) $a = 26$

10 Which of these equations does **not** have a solution of $m = 3$? [1]

(A) $6m = 18$ (B) $3m + 2 = 11$ (C) $6 - m = 3$ (D) $\frac{m}{3} = 6$

11 Solve the inequality $-6w \geq 12$. [1]

(A) $w \leq -2$ (B) $w \leq 6$ (C) $w \geq -2$ (D) $w \geq 6$

The lines p and q are drawn on the number plane and are used to answer questions 12 and 13.

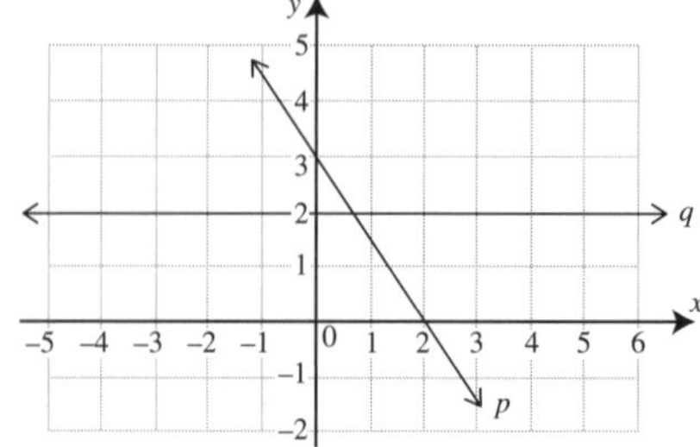

12 What is the y-intercept of the line p? [1]

(A) $-\frac{3}{2}$ (B) $\frac{3}{2}$ (C) 2 (D) 3

Marks

13 Which of these is the equation of the line q?

(A) $y = 2$ (B) $x = 2$ (C) $y = 3x + 2$ (D) $y = 2x + 3$ 1

14 Solve the equation $(x + 4)(x - 2) = 0$.

(A) $x = -4, 2$ (B) $x = 4, -2$ (C) $x = -4, -2$ (D) $x = 4, 2$ 1

15 If $y = kx$, what is the value of k when $y = 16$ and $x = 2$?

(A) $k = 2$ (B) $k = 4$ (C) $k = 8$ (D) $k = 32$ 1

16 Which of these is a Pythagorean triad?

(A) {11, 59, 60} (B) {11, 60, 61} (C) {11, 61, 62} (D) {11, 62, 63} 1

17 Which of these is closest to the area of the shape, to the nearest square centimetre?

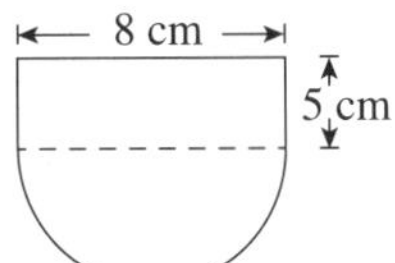

(A) 50 cm² (B) 65 cm² (C) 90 cm² (D) 101 cm² 1

18 Which of these is the surface area of a rectangular prism with sides 6 cm, 8 cm and 10 cm?

(A) 316 cm² (B) 348 cm² (C) 376 cm² (D) 480 cm² 1

19 Which of these is closest to $12 \times \cos 18°$?

(A) 11.41 (B) 11.42 (C) 17.60 (D) 17.61 1

The diagrams show two similar figures and are used to answer questions 20 and 21.

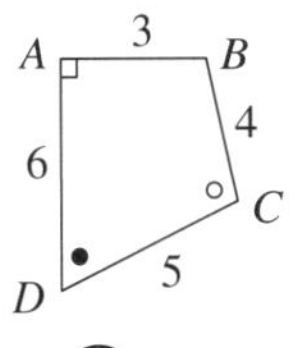

20 If $ABCD$ is the original and $PQRS$ is its image, what is the enlargement factor?

(A) 1.5 (B) 2 (C) 3 (D) 6 1

21 Which of these is the value of m?

(A) 4.5 (B) 6 (C) 6.5 (D) 7 1

22 For the scores 16, 18, 23, 25, 27 what is the interquartile range?

(A) 6 (B) 7 (C) 8 (D) 9 1

23 A set of scores is recorded in a grouped data table using the classes 16–20, 21–25, 26–30 and 31–35. Which of these is one of the class centres?

(A) 16 (B) 20.5 (C) 28 (D) 35 1

A box contains numbered balls. A ball is selected at random and then returned to the box. This process is repeated another 59 times and the results detailed in the relative frequency table. The table is used to answer questions 24 and 25.

Score	Frequency	Relative frequency
1	12	0.2
2	20	
3	9	
4		x
5	4	
Total	60	

24 Which of these is the value of x?

(A) 0.1 (B) 0.15 (C) 0.2 (D) 0.25 1

25 If only 40 draws occurred, what would be the most likely number of times a ball with the number 3 would be drawn?

(A) 2 (B) 4 (C) 6 (D) 8 1

Total marks

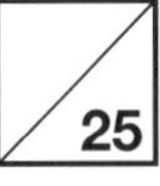

EXAM PAPER: LEVEL 1 — PART B

Marks

1 Rewrite 524.59 in scientific notation. ______ [1]

2 Christo scored 32 out of 40 for his trigonometry topic test. Express this result as a percentage. ______ [1]

3 A book has been priced at \$27. What will be the new price if the book was purchased at a 30%-discount sale? ______ [1]

4 Darlene is paid an hourly rate of \$19.30. What is she paid if she works 8 hours at double-time? ______ [1]

5 Last financial year Merrill's gross pay was \$84 350 and she had deductions totalling \$3540. What was her net pay? ______ [1]

6 Simplify $8y - 3z + 2z - 5y$. ______ [1]

7 Expand $5a(2a - 4)$. ______ [1]

8 Simplify $24x^6 \div 4x^2$. ______ [1]

9 Simplify $3a^2b^4 \times 6a^3b$. ______ [1]

10 Solve the equation $\frac{a}{3} = -4$. ______ [1]

11 If $M = \frac{p-q}{r}$, what is the value of M when $p = 12$, $q = 8$ and $r = 4$? ______ [1]

12 Write the inequality in terms of x that matches the graph:

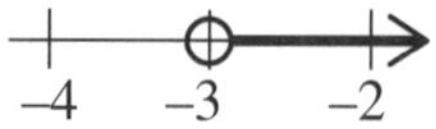

(number line: −4, −3, −2; open circle at −3, arrow pointing right)

______ [1]

13 Solve $\frac{y+5}{3} < 4$. ______ [1]

14 What is the gradient of the line with equation $y = 3 - 2x$? ______ [1]

15 Solve the equation $g^2 - 9 = 7$. ______ [1]

16 Factorise $x^2 + 7x - 18$. ______ [1]

17 If $P = 2.4n$, what is the value of n when $P = 12$? ______ [1]

18 The formula $C = 5.2n$ is used to determine the cost C, in dollars, of n m of garden edging. What will be the cost of 8.6 m of edging? ______ [1]

19 Evaluate $\sqrt{5^2 + 10^2}$, leaving your answer correct to two decimal places. ______ [1]

20 A cube has a side length of 10 cm. What is the surface area of the cube? ______ [1]

EXAM PAPER: LEVEL 1 — PART B

Marks

21 What is the volume of the triangular prism? ______ 1

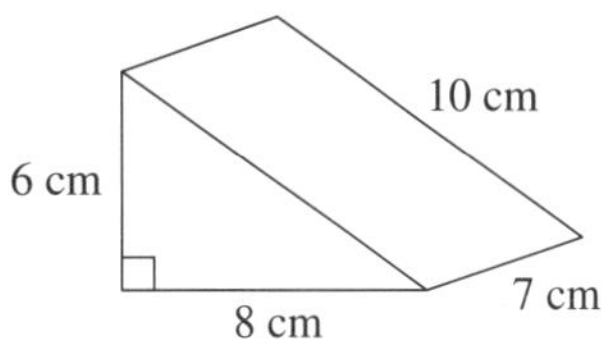

22 Write the expression for tan 37°. ______ 1

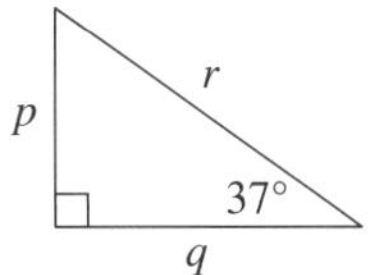

23 A map is drawn using a scale of 1 cm = 15 km. If the distance between two towns is 120 km, what is the distance between the two towns on the map? ______ 1

The diagram shows an original shape *WXYZ* and its image *MNPQ*, and is used to answer questions 24 to 26.

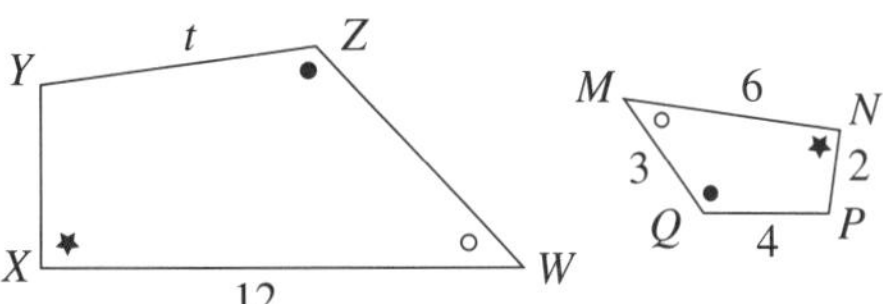

24 Complete: $\frac{MN}{WX}$ = ____ ______ 1

25 What angle matches *XYZ*? ______ 1

26 What is the value of *t*? ______ 1

27 The range of a set of scores is 28. If the lowest score is 67, what is the highest score? ______ 1

28 What is the median of the scores?

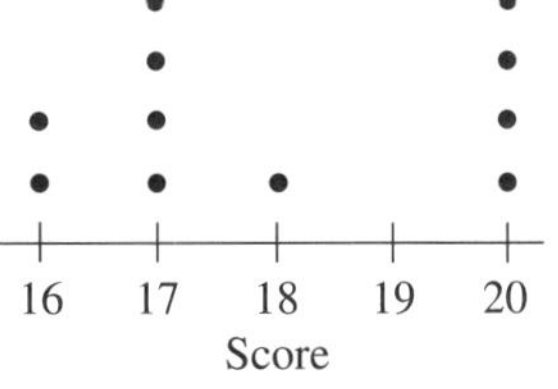

______ 1

29 Twenty coloured balls are placed in a box. Eight of the balls are purple, seven are pink and the remainder are red. If a ball is chosen at random, what is the probability that it is **not** red? ______ 1

30 A coin is tossed and the result is tails. The coin is tossed again and the result is tails again. What is the probability that the next toss will result in tails again? ______ 1

Total marks

EXAM PAPER : LEVEL 1 PART C

1 Complete the table: 6 marks

Fraction	Decimal	Percentage
$\frac{7}{25}$		
	0.05	
		$7\frac{1}{2}\%$

2 Sophie receives 4 weeks holiday loading at 17.5% of her normal pay each year. If her normal pay is $1630 per week, find the amount of holiday loading she will receive. 2 marks

3 Simplify $12a - 2a \times 3$. 2 marks

4 Simplify $(-3p)^2 - 2p \times 5p$. 2 marks

5 Simplify $4x^0 - (2x)^0$. 2 marks

6 Solve:

a $4g + 1 = 2g + 5$ 2 marks

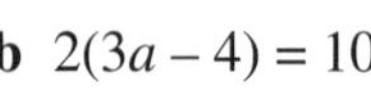

b $2(3a - 4) = 10$ 2 marks

7 If $T = \frac{m}{2}(a + b + c)$, what is the value of T when $a = 4$, $b = 7$, $c = 12$ and $m = 8$. 2 marks

8 Complete the table for $y = 3x - 2$ and graph the lines on the number plane. 3 marks

X	0	1	2
Y			

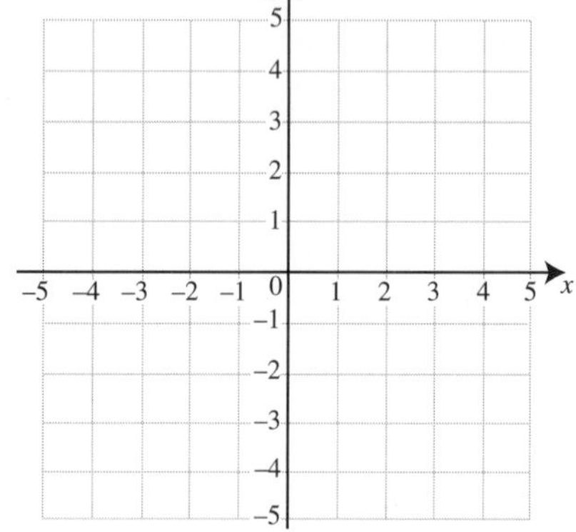

9 Expand and simplify $(2a - 5)(a + 3)$. 2 marks

10 Solve $p^2 - 6p + 8 = 0$. 2 marks

11 Kristin uses two eggs to make a dozen muffins.

a Complete the formula linking the number of muffins (m) with the number of eggs (e).

$m =$ ____________ 1 mark

b Use the formula to find the number of muffins that can be made with five eggs. 1 mark

12 Find the value of the pronumeral, correct to three decimal places. 2 marks

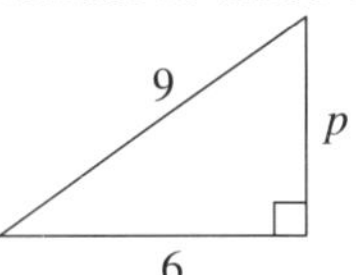

13

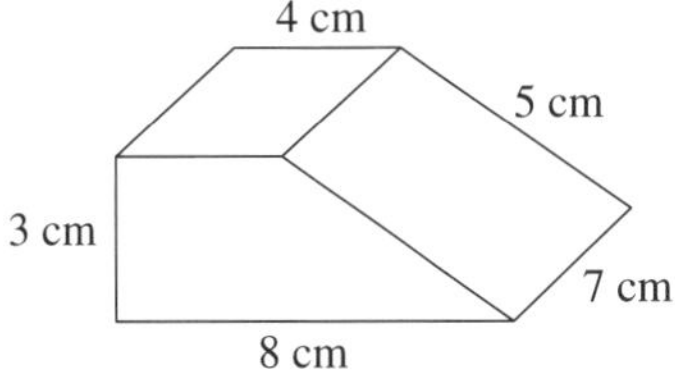

Find the volume of the prism. 3 marks

14

8.6
53°
p

What is the value of p, correct to two decimal places? 2 marks

15

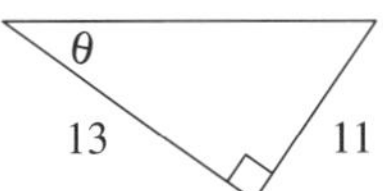

Find the size of the angle θ, to the nearest degree. 2 marks

16

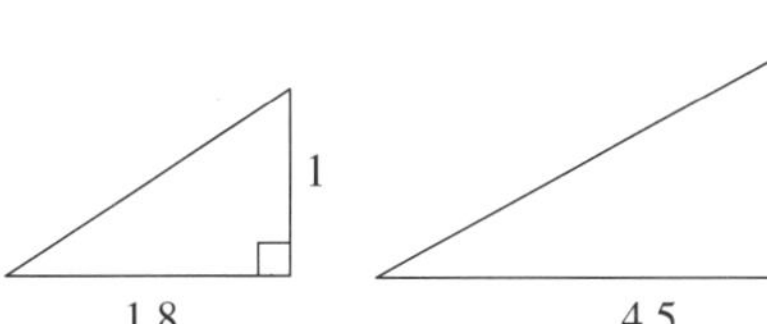

Late one afternoon, a 1-m high stick cast a 1.8-m shadow. At the same instant, a fence post cast a shadow 4.5 m long. How high was the fence post? 2 marks

17 The pulse rates of two students are recorded at particular times in the day and the results recorded in the stem-and-leaf display.

Heart rate (per minute)

Tom		**Darcie**
74	6	368
9864	7	006
71	8	14

a What was the range of Tom's rates? 1 mark

b What was the median of Darcie's rates? 1 mark

18 A group of students were surveyed to find who owned a bike and a skateboard. The results are shown in the Venn diagram.

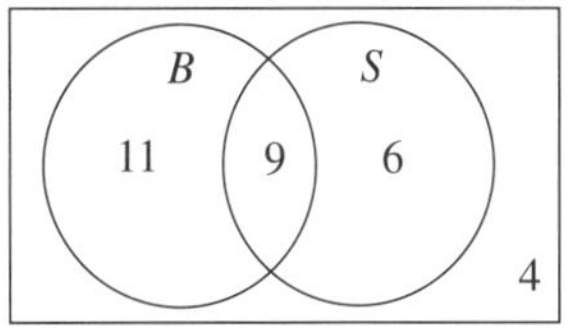

B: Bike
S: Skateboard

a How many students were surveyed? 1 mark

b If a student is chosen at random, what is the probability that they:

i owned both a bike and a skateboard? 1 mark

ii did **not** own a skateboard? 1 mark

Total marks for PART C

Total marks

Exam Paper: Level 2

Instructions for all parts
- Attempt all questions.
- Calculators are allowed.

Time allowed: $1\frac{1}{2}$ hours

Part A: Allow about 20 minutes for this part.
Part B: Allow about 30 minutes for this part.
Part C: Allow about 40 minutes for this part.

Total marks: 100

EXAM PAPER: LEVEL 2 — PART A

Fill in only one circle for each question.

Marks

1 Which of these is 8765.95, written in scientific notation, correct to three significant figures?

Ⓐ 8.766×10^3 Ⓑ 8.767×10^3 Ⓒ 8.76×10^3 Ⓓ 8.77×10^3 [1]

2 Which of these is the best buy?

Ⓐ \$4.05 for 300 mL Ⓑ \$6.90 for 500 mL Ⓒ \$9.90 for 750 mL Ⓓ \$12.60 for 940 mL [1]

3 Audrey works from 8:15 am to 12:00 pm in her casual employment and is paid \$19.20 per hour. Find her pay.

Ⓐ \$66.24 Ⓑ \$72.00 Ⓒ \$79.68 Ⓓ \$81.60 [1]

4 Mia received a 15% discount on the price of a rug. If she paid \$289 for the rug, what was the original price?

Ⓐ \$304 Ⓑ \$328 Ⓒ \$332.35 Ⓓ \$340 [1]

5 Expand and simplify $3(y - 2) + 2(y - 5)$.

Ⓐ $5y - 7$ Ⓑ $5y - 11$ Ⓒ $5y - 15$ Ⓓ $5y - 16$ [1]

6 Factorise $4a^2 - 2a$.

Ⓐ $a(a - 2)$ Ⓑ $4a(a - 2)$ Ⓒ $2a(2a - 1)$ Ⓓ $2(2a - 1)$ [1]

7 Which of the following is equal to 0.0607 written in scientific notation?

Ⓐ 6.07×10^{-2} Ⓑ 6.7×10^{-2} Ⓒ 6.07×10^{-3} Ⓓ 6.07×10^{-4} [1]

8 Simplify $16m^2 \div 4m^{-3}$.

Ⓐ $4m^{-1}$ Ⓑ $12m^{-1}$ Ⓒ $4m^5$ Ⓓ $12m^{-5}$ [1]

9 What is the value of m when $n = 5$ is substituted in $2m - n = 7$?

Ⓐ $m = 2$ Ⓑ $m = 6$ Ⓒ $m = 8$ Ⓓ $m = 12$ [1]

10 Which of these is the solution of the equation $2 - (2t + 1) = 3$?

Ⓐ $t = -1$ Ⓑ $t = 0$ Ⓒ $t = 1$ Ⓓ $t = 2$ [1]

11 Which of these is the solution of the inequality $1 - y > 3$?

Ⓐ 

Ⓑ 

Ⓒ 

Ⓓ 

[1]

12 Which of these points does the line $3x - 2y - 4 = 0$ pass through?

Ⓐ (2, 1) Ⓑ (0, 2) Ⓒ (1, 3) Ⓓ (1, 0) [1]

13 Which of these is the solution of $m^2 + m - 6 = 0$?

Ⓐ $m = 6, -1$ Ⓑ $m = -6, 1$ Ⓒ $m = -3, 2$ Ⓓ $m = 3, -2$ [1]

14 If T is directly proportional to n and when $T = 24$, $n = 4.8$, what is the constant of proportionality?

Ⓐ 5 Ⓑ 19.2 Ⓒ 28.8 Ⓓ 115.2 [1]

15 Which of the following represents a direct proportion?

Ⓐ $P = \frac{2}{n}$ Ⓑ $P = \frac{2}{n + 3}$ Ⓒ $P = \sqrt{n}$ Ⓓ $P = \frac{n}{2}$ [1]

Marks

16 Which of these is the length of a diagonal of a rectangular television screen with sides 63 cm and 84 cm?

Ⓐ 102 cm Ⓑ 103 cm Ⓒ 104 cm Ⓓ 105 cm 1

17 Which of these is the area of a right triangle with longer sides 12 cm and 13 cm?

Ⓐ 25 cm^2 Ⓑ 30 cm^2 Ⓒ 45 cm^2 Ⓓ 78 cm^2 1

The diagram shows a closed cylinder and is used to answer questions 18 and 19.

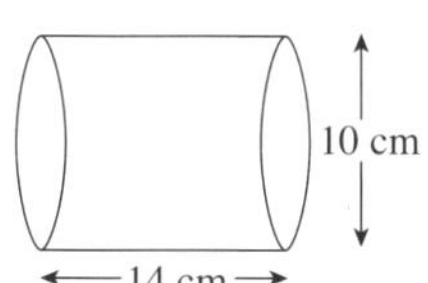

18 Which of these is closest to the surface area of the cylinder?

Ⓐ 440 cm^2 Ⓑ 597 cm^2 Ⓒ 880 cm^2 Ⓓ 1508 cm^2 1

19 Which of these is the exact volume of the cylinder?

Ⓐ 140π cm^3 Ⓑ 350π cm^3 Ⓒ 1400π cm^3 Ⓓ 1960π cm^3 1

20 Which of these is closest to the value of θ if $\tan \theta = \frac{4.2}{5.9}$, given θ is acute?

Ⓐ 35°26′ Ⓑ 35°27′ Ⓒ 35°44′ Ⓓ 35°45′ 

21 The diagram shows a pair of similar triangles. Which of these is closest to the value of q?

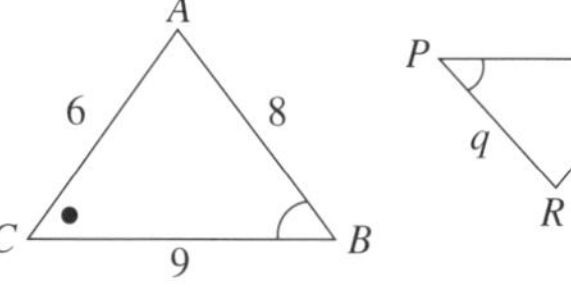

Ⓐ 5.3 Ⓑ 5.5 Ⓒ 5.7 Ⓓ 5.8 1

22 The ages of people at a concert were recorded in the box plot opposite.

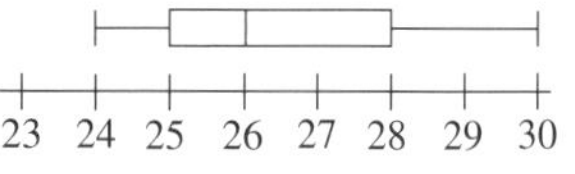

If there were 120 people at the concert, how many were 26 or over?

Ⓐ 4 Ⓑ 30 Ⓒ 40 Ⓓ 60 1

23 The diagram shows a cumulative frequency table for a set of scores.

Score	Frequency	Cumulative frequency
34	6	
35	14	
36	11	
37	9	t
Total		

Which of these is the value of t?

Ⓐ 4 Ⓑ 9 Ⓒ 40 Ⓓ 333 1

24 Sam has found that the probability that his dog barks when he arrives home from school is 0.3. What is the probability that his dog barks on two consecutive school days?

Ⓐ 0.09 Ⓑ 0.33 Ⓒ 0.6 Ⓓ 0.9 1

25 A survey of retirement-village residents found the number of people who hold a driver's licence.

Driver's licence	Under 85	85 or over
yes	13	11
no	10	16

If a resident who has a licence was chosen at random, what is the probability that they were aged under 85?

Ⓐ $\frac{13}{50}$ Ⓑ $\frac{23}{50}$ Ⓒ $\frac{13}{24}$ Ⓓ $\frac{13}{23}$ 1

Total marks /25

EXAM PAPER : LEVEL 2 — PART B

Marks

1 Express 13 804 000 in scientific notation, correct to four significant figures. ____________ 1

2 Jacob's mass is 62 kg. Over his vacation he increased his mass by 1.5% and then, once back at work, reduced his mass by 2%. What is his new mass, correct to one decimal place? ____________ 1

3 Find the amount of simple interest on \$1350 at 6% p.a. for 8 years. ____________ 1

4 Expand and simplify $(2c + 1)(c - 3)$. ____________ 1

5 Simplify $\frac{4}{3p} + \frac{2}{3p}$. ____________ 1

6 Expand $4w^3(3w^2 - 5y)$. ____________ 1

7 If $a = 3$ and $b = 4$, find the value of $2a^0 - b^0$. ____________ 1

8 What is the value of $64^{-\frac{1}{2}}$? ____________ 1

9 Solve the equation $\frac{-1}{q} = 2$. ____________ 1

10 Solve $2x^2 = 8$. ____________ 1

11 Find the value of b if $a = -4$ in $b = 3 - 2a$. ____________ 1

12 What is the solution to the equation $4 - 2x > -6$. ____________ 1

13 What is the point of intersection of the lines $y = 2x + 1$ and $x = 1$? ____________ 1

14

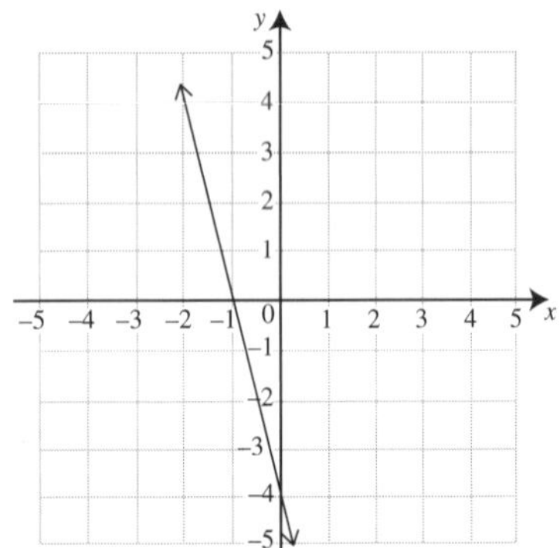

What is the gradient of the line? ____________ 1

15 What is the equation of the line passing through (–2, 4) and parallel to the x-axis? ____________ 1

16 Solve the equation $a^2 - 4a - 21 = 0$. ____________ 1

17 Factorise $4x^2 - 4x - 8$. ____________ 1

18 What is the exact length of the hypotenuse of a right triangle with shorter sides 9 cm and 8 cm? ____________ 1

19

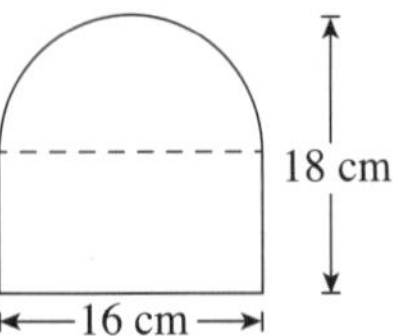

What is the area of the shape, correct to two decimal places? ____________ 1

EXAM PAPER: LEVEL 2 — PART B

Marks

20 The curved surface area of a cylinder is 80π cm^2. If the height of the cylinder is 8 cm, what is the volume of the cylinder, correct to two decimal places? ______ 1

21

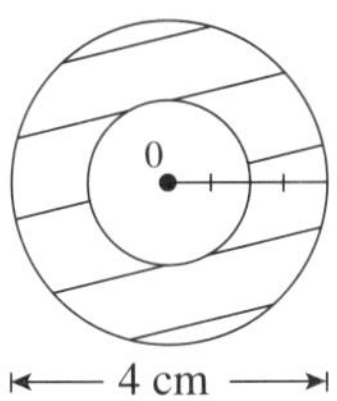

Find the area of the shaded region, in terms of π.

______ 1

22

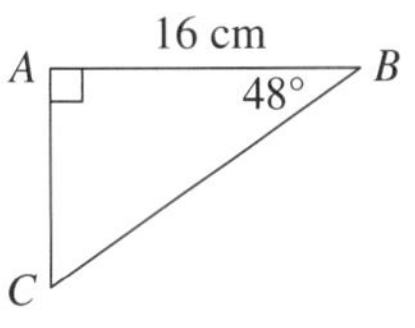

Find the length of AC, correct to two decimal places.

______ 1

23

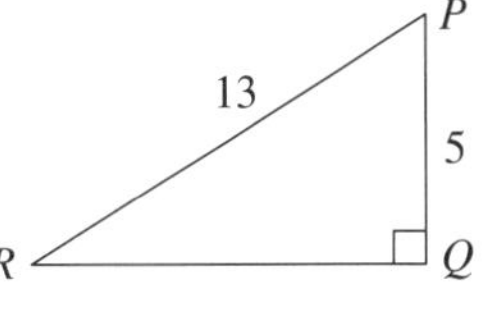

What is the value of sin P?

______ 1

24

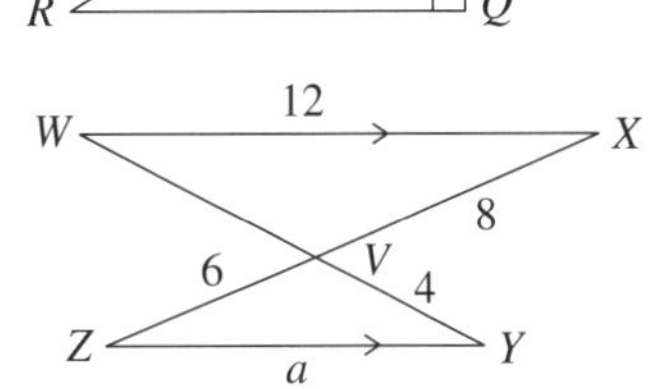

Find the value of a.

______ 1

The scores are used to answer questions 25 to 27. 11, 17, 13, 15, 18, 14, 14, 12

25 What is the median of the scores? ______ 1

26 What is the lower quartile of the scores? ______ 1

27 Draw a box plot for the scores.

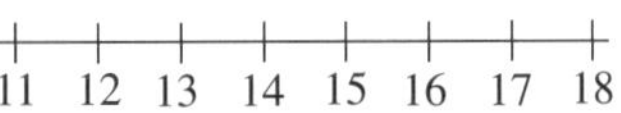

1

28 Sarah has two boxes. In Box A there are three white balls and two black balls. In Box B there are two white balls and three black balls. What is the probability that when she chooses a ball from each box they are both black? ______ 1

A bag contains three discs: a red disc, a green disc and a blue disc. Two discs are selected at random from the bag. This information is used to answer questions 29 and 30.

29

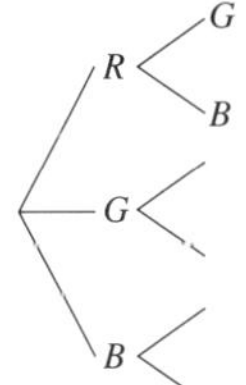

Use the tree diagram to list the outcomes.

______ 1

30 What is the probability of selecting a green ball? ______ 1

Total marks /30

EXAM PAPER: LEVEL 2 — PART C

1 The tax table is used to determine the amount of tax to be paid.

Taxable income	Tax on this income
\$0–\$18 200	Nil
\$18 201–\$37 000	19c for each \$1 over \$18 200
\$37 000–\$87 000	\$3572 plus 32.5c for each \$1 over \$37 000
\$87 000–\$180 000	\$19 822 plus 37c for each dollar over \$87 000
\$180 001 and above	\$54 232 plus 47c for each \$1 over \$180 000

Find the tax payable on a taxable income of \$83 000. 2 marks

2 Find the simple interest on \$12 600 at 5.75% p.a. for 8 months. 3 marks

3 Expand and simplify $3a(2a-5)-a(a-1)$. 2 marks

4 Simplify $\frac{3a-4}{3}+\frac{5a}{2}$. 2 marks

5 Find the exact value of $\left(\frac{9}{64}\right)^{-\frac{1}{2}}$. 2 marks

6 Solve:

a $3(4m+5)-2(m-1)=-3$ 3 marks

b $\frac{2w+5}{3}=\frac{3w}{2}$ 3 marks

7 Solve simultaneously: 3 marks

$2a+3b=1$ $\qquad$ $2a-b=5$

8 Express t as the subject of $m=s-ut$. 2 marks

9 Simplify $\frac{ab-a}{b^2+b-2}$. 2 marks

10 The points $A(-3,-1)$ and $B(5, 3)$ are plotted on the number plane.

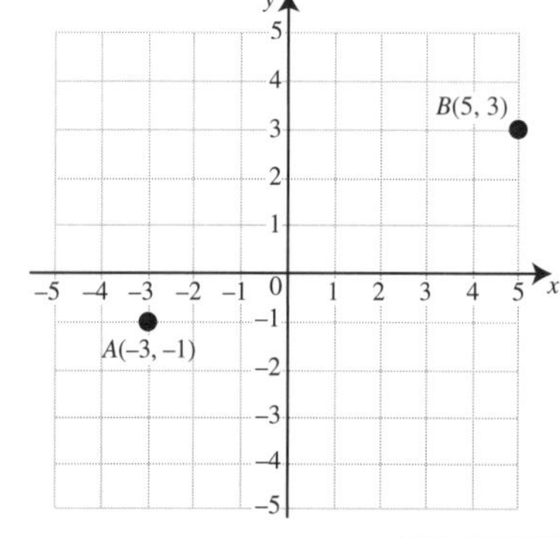

Find:

a length of AB 2 marks

b gradient of AB 2 marks

11 P is directly proportional to the square of n and when $n = 5$, $P = 90$.

a Write a formula linking P and n and find the constant of proportionality. 2 marks

b What is the value of P when $n = 8$? 2 marks

12

2 7

5 x

Find the exact value of x. 3 marks

13

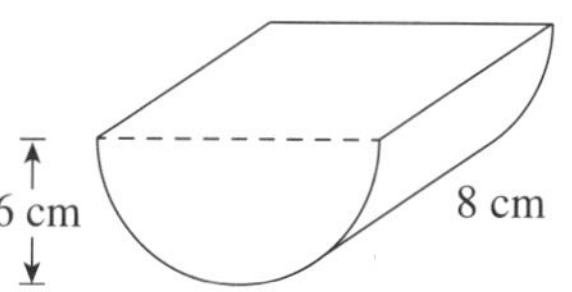

What is the volume of the solid, correct to three significant figures? 3 marks

14

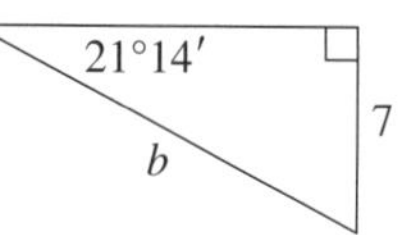

Find the value of b, correct to two significant figures. 2 marks

15

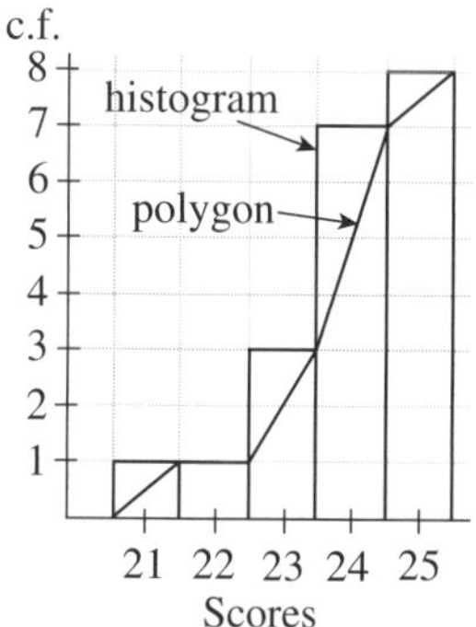

Using the cumulative frequency histogram and polygon, what is the:

a mode of the scores? 1 mark

b median of the scores? 1 mark

16 A drawer contains two white socks and two black socks. Two socks are chosen at random. What is the probability of choosing:

a two black socks? 1 mark

b matching socks? 1 mark

c different socks? 1 mark

Total marks for PART C 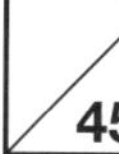 45 **Total marks**  100

Exam Paper: Level 3

Instructions for all parts
- Attempt all questions.

Time allowed: $1\frac{1}{2}$ hours
- Calculators are allowed.

Part A: Allow about 20 minutes for this part.
Part B: Allow about 30 minutes for this part.
Part C: Allow about 40 minutes for this part.

Total marks: 100

EXAM PAPER: LEVEL 3 — PART A

Fill in only one circle for each question.

Marks

1 Evaluate $\dfrac{8.5471}{\sqrt{12.6-3.12}}$, correct to four significant figures.

(A) 0.7121 (B) 0.7122 (C) 2.776 (D) 2.777 — 1

2 Which of these is the number of years \$500 will be invested at 4.5% p.a. simple interest to earn \$135?

(A) 5 (B) 6 (C) 8 (D) 9 — 1

3 Ethan invests \$1000 in an account offering 4% p.a. compound interest. In how many years will he double his money?

(A) 2 (B) 18 (C) 20 (D) 25 — 1

4 In one week Michael worked 24 hours normal time and 6 hours at time-and-a-half. If Michael received \$805.20, how much will he be paid to work 20 hours normal time and 4 hours at double-time?

(A) \$639.90 (B) \$642.80 (C) \$644.16 (D) \$683.20 — 1

5 Which of these is the difference between $4m - 2n$ and $n - 3m$?

(A) $3m - 5n$ (B) $m - 3n$ (C) $7m - n$ (D) $7m - 3n$ — 1

6 Factorise $ap - 3aq - 2bp + 6bq$.

(A) $(a - 3b)(p - 2q)$ (B) $(a - 2b)(p - 3q)$ (C) $(a + 3b)(p - 2q)$ (D) $(a - 3b)(p + 2q)$ — 1

7 Which of these is equivalent to $\dfrac{5}{\sqrt{2m+3}}$?

(A) $(2m + 3)^{\frac{5}{2}}$ (B) $(2m + 3)^{-\frac{5}{2}}$ (C) $5(2m + 3)^{-\frac{1}{2}}$ (D) $5(2m + 3)^{\frac{1}{2}}$ — 1

8 Simplify $16^x \div 4^{-2x}$.

(A) 1 (B) 4^{-x} (C) 4^{-2x} (D) 4^{4x} — 1

9 Which of these pairs of equations has solutions $p = 3$ and $q = 1$?

(A) $p + q = 4$ and $p - q = -2$
(B) $3p - q = 8$ and $2p + q = 4$
(C) $2p + 3q = 9$ and $2p - q = 5$
(D) $p - 2q = -1$ and $p + 2q = 5$ — 1

10 Solve the equation $\dfrac{3b}{2} - \dfrac{b}{4} = -5$.

(A) $b = -8$ (B) $b = -4$ (C) $b = -2$ (D) $b = 4$ — 1

11 Which of these is the gradient of the line with equation $6x + 3y + 4 = 0$?

(A) -2 (B) $-\frac{1}{2}$ (C) $\frac{1}{2}$ (D) 2 — 1

Marks

12 The line $y = ax - 4$ passes through the point (3, 2) and cuts the axes at P and Q. Which of these is the length of PQ?

(A) $\sqrt{12}$ units (B) 4 units (C) $\sqrt{20}$ units (D) 5 units — 1

13 The point $P(3, -4)$ lies on a circle with its centre at the origin. Which of these points also lies on the circle?

(A) (5, 0) (B) (5, 5) (C) $(\sqrt{20}, 5)$ (D) $(\sqrt{12}, -8)$ — 1

14 T is inversely proportional to the square of n, and $n = 0.5$ when $T = 6$. Which of these is the equation which links T and n?

(A) $T = \frac{3}{n^2}$ (B) $T = \frac{3}{2n^2}$ (C) $T = \frac{12}{n^2}$ (D) $T = \frac{24}{n^2}$ — 1

15 Two poles are 8 m apart and are 6 m and 11 m tall. Which of these is closest to the distance between the tops of the poles?

(A) 9.4 m (B) 9.5 m (C) 9.6 m (D) 9.7 m — 1

16 A rectangular prism with a capacity of 560 mL has a length of 10 cm and width of 8 cm. Which of these is the surface area of the prism?

(A) 412 cm^2 (B) 420 cm^2 (C) 436 cm^2 (D) 448 cm^2 — 1

17 A rectangle with an area of 168 cm^2 is bent to form a cylinder 14 cm high. Which of these is closest to the volume of the cylinder?

(A) 160 cm^3 (B) 476 cm^3 (C) 1583 cm^3 (D) 2352 cm^3 — 1

18 Which of these is the expression for cos 55°?

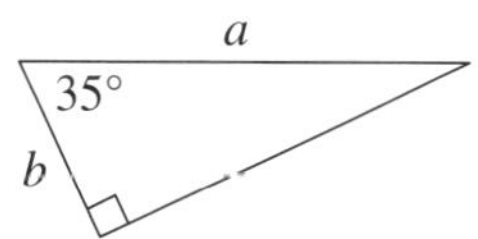

(A) $\frac{b}{a}$ (B) $\frac{b}{\sqrt{a^2 - b^2}}$ (C) $\frac{\sqrt{a^2 - b^2}}{a}$ (D) $\frac{\sqrt{a^2 - b^2}}{b}$ — 1

The diagram is used to answer questions 19 and 20.

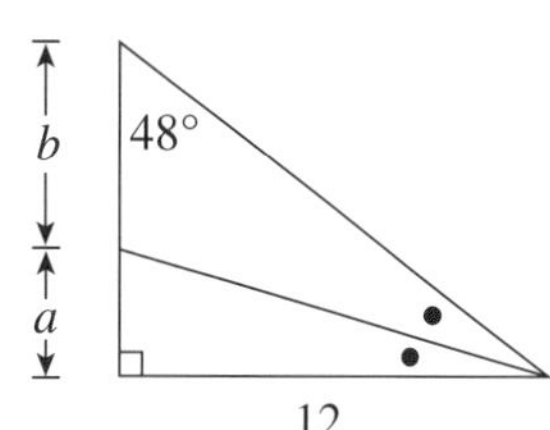

19 Which of these is the value of a, correct to two decimal places?

(A) 4.61 (B) 5.34 (C) 5.68 (D) 6.66 — 1

20 Which of these is the value of b, correct to two significant figures?

(A) 4.5 (B) 5.3 (C) 6.2 (D) 6.7 — 1

Marks

21 Which of these is the value of x?

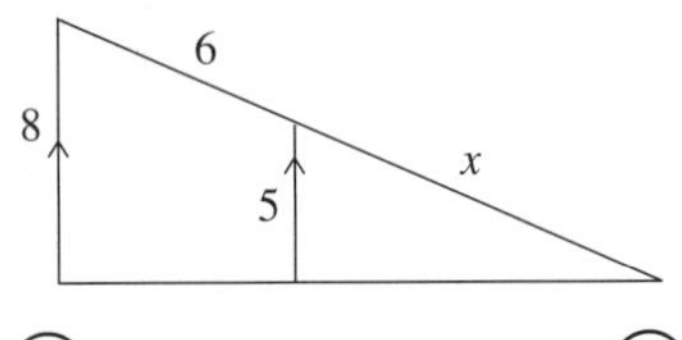

(A) 3.75 (B) 7.5 (C) 8.5 (D) 10 1

The results in two essays for a certain class of 28 students are recorded in the box plots below and are used to answer questions 22 and 23.

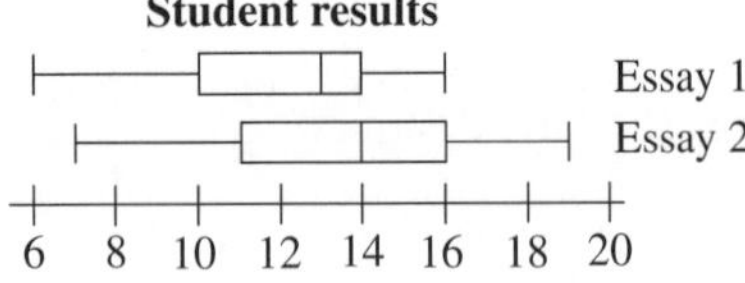

22 How many essays scored at least 14 marks?

(A) 10 (B) 11 (C) 14 (D) 21 1

23 Which statement is **not** correct?

(A) The median for essay 1 is 1 less than the median for essay 2.

(B) The range for essay 1 is 2 less than the range for essay 2.

(C) The maximum mark for essay 1 is the same as the upper quartile for essay 2.

(D) The IQR for essay 1 is 2 less than the IQR for essay 2. 1

24 A batch of 600 items is examined. The probability that an item from this batch is defective is 0.04. How many items from this batch are **not** defective?

(A) 24 (B) 240 (C) 360 (D) 576 1

25 Two identical biased coins are tossed together and the outcome is recorded. After a large number of trials it is observed that the probability that both coins land showing tails is 0.64. What is the probability that both coins land showing heads?

(A) 0.04 (B) 0.08 (C) 0.32 (D) 0.36 1

Total marks /25

EXAM PAPER: LEVEL 3 PART B

1 What is the middle of $\frac{2}{5}$ and $1\frac{1}{2}$? ______

2 Evaluate $\sqrt{\frac{49.87}{12.1^2 - 31.098}}$, correct to three significant figures. ______ 

3 What is the amount of compound interest earned on an investment of \$4500 at 5% p.a. for 3 years? ______ 

4 A painting was purchased in May 2014 for \$85 000. If in the following two years it depreciated at an annual rate of 6%, what was the value of the painting in May 2016? ______ 

5 Simplify $2 - \frac{x - y}{x}$. ______ 

6 Simplify, expressing the answer with positive indices $(27y^{-6})^{-\frac{1}{3}}$. ______ 

7 Rewrite 0.000 0317 in scientific notation. ______ 1

8 Factorise $2^{a+b} - 2^a$. ______ 1

9 Solve $\frac{3}{2x} = -\frac{5}{6}$. ______ 1

10 Solve $\sqrt{6 - 2x} = 2$. ______ 

11 Solve the simultaneous equations $n = 3m$ and $n = 2m - 3$. ______ 

12 If $C = 2\pi r$ and $A = \pi r^2$, express A in terms of C. ______ 

13 What is the solution of the inequality $1 - x > 3 + x$? ______ 

14 $M(1, 2)$ is the midpoint of XY, where $X(-4, 6)$. What are the coordinates of Y? ______ 

15 Expand and simplify $(a - b)^2 - (a + b)^2$. ______ 

16 Factorise $6 - 4x - 2x^2$. ______ 

17 If a bag of lollies is divided among 24 children, they will each receive three lollies. How many will each receive if the number of children is reduced by 6? ______ 

18 A runner leaves P and runs 2.8 km east and then 3.2 km north. She then turns and runs back to P. What is her total running distance, to the nearest metre? ______ 

The diagram shows a triangular prism of volume 960 cm³ and is used to answer questions 19 and 20.

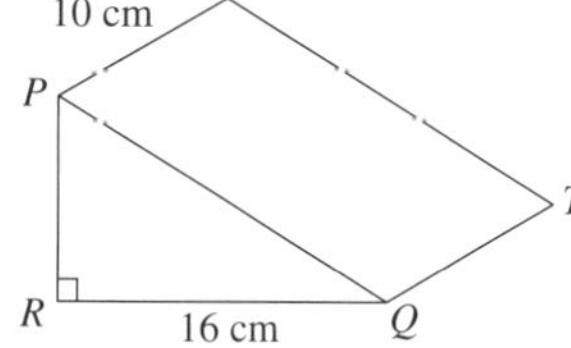

19 What is the length of PR? ______ 1

20 Find the surface area of the prism. ______ 1

Marks

21 If the bearing of P from Q is 245°, what is the bearing of Q from P? __________ 1

22 Find the size of θ, to the nearest minute. __________ 1

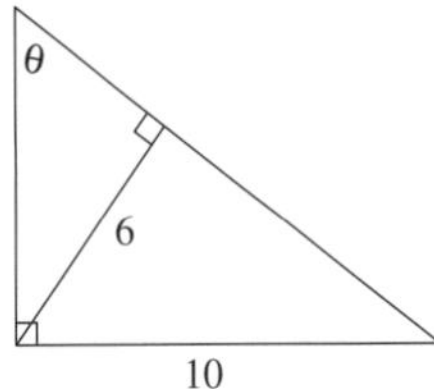

23 Find the perimeter of the triangle, correct to two decimal places. __________ 1

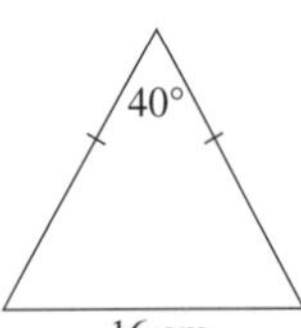

The diagram is used to answer questions 24 and 25.

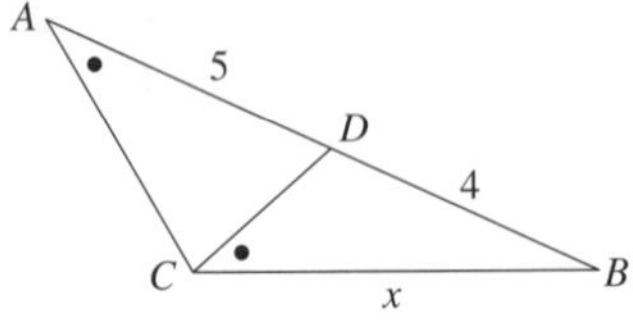

24 Name the triangle that is similar to ABC. __________ 1

25 What is the value of x? __________ 1

The stem-and-leaf display shows maximum temperatures during two fortnights and is to be used to answer questions 26 and 27.

Maximum temperatures °C

January		February
	1	9
998754	2	34578
98742	3	012a469
321	4	1

26 What is the value of a if the modal temperature in January is 5 degrees less than the modal temperature in February? __________ 1

27 What is the difference in the two medians? __________ 1

28 Two ordinary dice are rolled. The score is the sum of the numbers on the top faces. What is the probability that the score is **not** 4? __________ 1

29 In a game, a turn involves rolling two dice, each with faces marked 0, 1, 2, 3, 4 and 5. The score is calculated by multiplying the two numbers uppermost on the dice. What is the probability of scoring zero? __________ 1

30 A group of 28 students are surveyed regarding the way they check weather forecasts. Eighteen said they use their phone while 15 said they use television. Five students reported that they use neither source. By completing the Venn diagram, what is the probability that a randomly chosen student uses both sources? __________ 1

P T

P: Phone
T: Television

Total marks

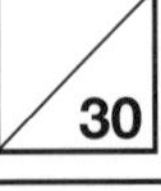

EXAM PAPER: LEVEL 3 PART C

1 Express $0.7\dot{3}$ in the form of $\frac{p}{q}$ where p and q are integers. [2 marks]

2 Evaluate $\left(\frac{1.89}{0.007 \times 0.0314}\right)^2$, leaving your answer in scientific notation, correct to two significant figures. [2 marks]

3 A caravan with a price of \$84 000 was purchased by paying a deposit of 15% and 6% p.a. simple interest on the balance over 4 years. Find the amount of:

a the deposit [1 mark]

b the interest [1 mark]

c each monthly instalment [1 mark]

4 Expand and simplify $(2x + 3y)^2 \quad (2x - 3y)^2$. [3 marks]

5 Find the value of y if $3^{2y-1} = 27$. [2 marks]

6 Simplify $\left(\frac{40a^5b^3}{5a^2b^9}\right)^{-\frac{1}{3}}$. [2 marks]

7 Solve $\frac{2k-1}{3} - \frac{9-k}{4} = 7 - k$. [3 marks]

8 Solve: $3p - q = 14 \qquad 2p + 3q = -9$ [3 marks]

9 Find the equation of the line which passes through $(2, 5)$ and is parallel to $3x - y + 4 = 0$. [3 marks]

10 Solve $6m^2 - 5m - 6 = 0$. [2 marks]

11 On a farm the number of days food will last is inversely proportional to the number of cattle. A farmer has enough food to feed 20 head of cattle for 6 days.

a By finding the constant of proportionality, write an equation that links the number of days (d) that food will last for the number of cattle (n). [2 marks]

b How long will the food last if another 10 cattle are added to the herd? 1 mark

12 If $CB = 18$ cm and the circumference of the circular base is 24π cm, what is the height CM, correct to three significant figures? 3 marks

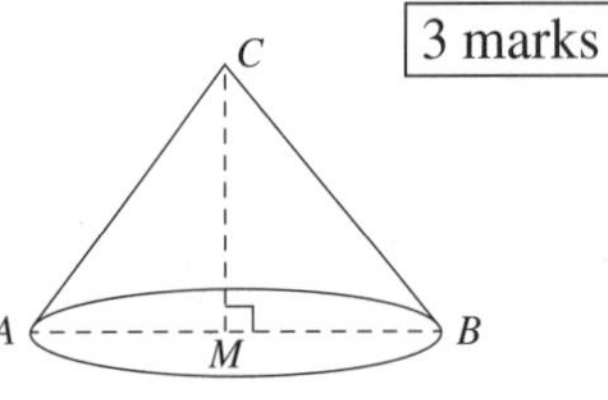

13 Find the surface area, correct to two decimal places. 3 marks

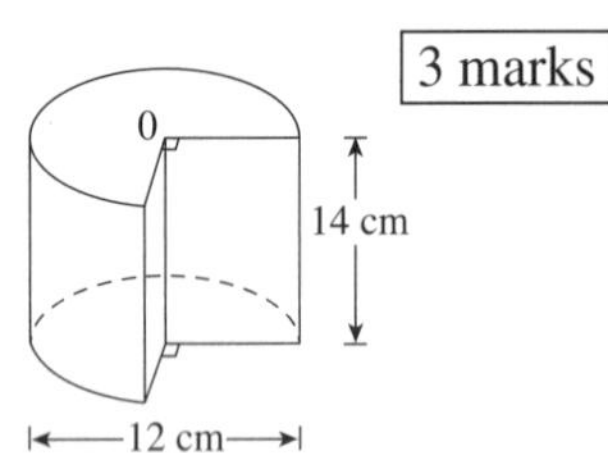

14 a Find the area of triangle STQ. 2 marks

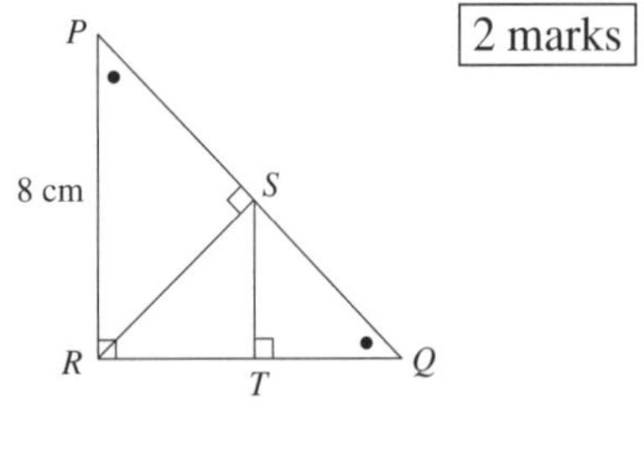

b Find the length of ST. 1 mark

15 Find the size of θ, correct to the nearest minute. 3 marks

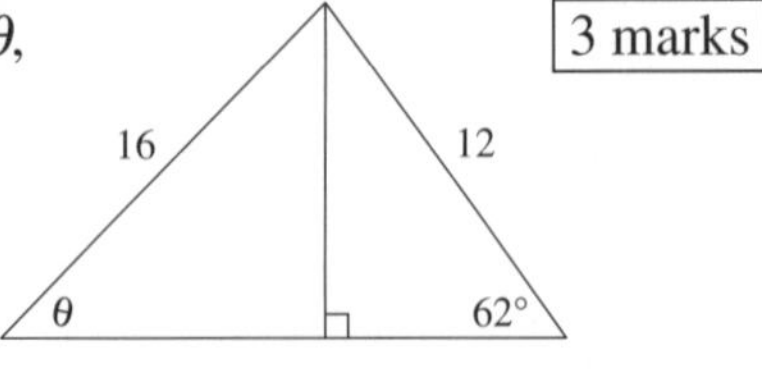

16 The box plot shows the delay times for the first 40 trains passing through a station on a given day.

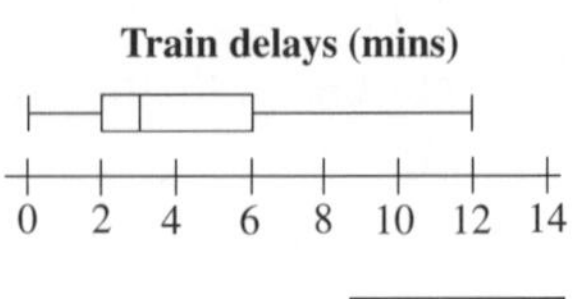

a Comment on the skewness of the data. 1 mark

b Due to a breakdown, the next eight trains were delayed for 10 minutes each. Explain the impact on the existing box plot. 2 marks

17 A bag contains five cards numbered 1 to 5. Two cards are chosen. What is the probability that:

a a 5 is chosen? 1 mark

b the product of the card numbers is greater than 15? 1 mark

Total marks for PART C  45

Total marks  100

Answers

Fractions, Decimals, Percentages and Ratios

Page 1 **1** B **2** A **3** A **4** D **5** D **6** B **7** C **8** B **9** C **10** D

Page 2 **1** C **2** C **3** D **4** D **5** D **6** B **7** B **8** A **9** A **10** D

Page 3 **1** B **2** C **3** D **4** A **5** D **6** C **7** C **8** B **9** A **10** A **11** D **12** A **13** C

Page 4 **14 a** 21.15 **b** 5.89 **c** 1.71 **15 a** 1 : 10 **b** 5 : 3 **c** 4 : 9 **16 a** 0.75 **b** 0.18 **c** 0.04 **17 a** 8% **b** 80% **c** 5% **18 a** $\frac{3}{5}$ **b** $\frac{1}{50}$ **c** $\frac{6}{25}$ **19 a** 8.6×10^2 **b** 4.59×10^3 **c** 3.7×10^6 **20 a** 267.5 **b** 5793.29 **c** 72 982 **21 a** 62 **b** 7 **22 a** 549 **b** 124 **23 a** 80% **b** 85%

Page 5 **1** C **2** A **3** D **4** A **5** A **6** B **7** D **8** A **9** C **10** A **11** D **12** D **13** A

Page 6 **14 a** 1.8 **b** 1.3 **c** 0.45 **15 a** 1.27×10^3 **b** 2.3908×10^4 **c** 2.506×10^6 **16 a** 20 : 1 **b** 1 : 9 **c** 1 : 200 **d** 1000 : 1 **17 a** \$288, \$432 **b** \$320, \$240, \$160 **18 a** 3^6 **b** 5^5 **c** 2^6 **19 a** 4.02 kg (2 dec. pl.) **b** 42 km/h **20 a** 36 km/h **b** 22.22 m/s (2 dec. pl.) **21 a** 40° **b** 0.504 kL

Page 7 **1** B **2** C **3** D **4** A **5** C **6** B **7** C **8** A **9** D **10** A **11** A **12** D **13** A

Page 8 **14 a** 1.08×10^2 **b** 6.29×10^3 **c** 1.08×10^9 **d** 5.25×10^{34} **15 a** $\frac{23}{33}$ **b** $\frac{4}{15}$ **16 a** 44% **b** 8% **17 a** $\frac{4}{15}$ **b** $\frac{19}{20}$ **c** 0.1 **18 a** 1.40 **b** 2.67 **c** 0.397 **19 a** 1160 m^2 per person **b** 24 **c** 8.33 minutes

Financial Mathematics

Page 9 **1** D **2** B **3** C **4** B **5** B **6** C **7** A **8** B **9** A **10** C

Page 10 **1** C **2** D **3** D **4** D **5** A **6** C **7** B **8** A **9** B **10** B

Page 11 **1** C **2** C **3** B **4** B **5** B **6** B **7** C **8** D **9** B **10** D **11** B **12** A **13** A

Page 12 **14 a** \$675.50 **b** \$347.40 **c** \$193 **15 a** \$12 150 **b** \$6750 **16 a** \$63 778, \$3985 **b** \$8071, \$90 360 **17 a** \$72 **b** \$8750 **18 a** \$315 **b** \$3780 **19 a** \$728 **b** \$4888 **20** \$5593

Page 13 **1** B **2** D **3** B **4** A **5** A **6** C **7** D **8** B **9** B **10** D **11** A **12** C **13** C

Page 14 **14 a** \$655.50 **b** \$3865.15 **c** \$25 613.24 **15 a** \$75 982 **b** \$16 241.15 **16 a** \$96, \$544 **b** 20%, \$400 **c** \$108, 15% **d** \$240, \$180 **e** \$400, 10% **17 a** \$3087 **b** \$658.56 **18 a** \$14 080 **b** 5%

Page 15 **1** D **2** C **3** D **4** C **5** B **6** C **7** B **8** A **9** B **10** B **11** D **12** B **13** D

Page 16 **14 a** \$85 600 **b** \$142 000 **15 a** \$7601.84 **b** \$1545 **16 a** \$1450.95 **b** \$1551.38 **17 a** \$2545 **b** \$660 **18 a** \$24 377.76 **b** \$1157.94 **19** 6%

Algebraic Techniques

Page 17 **1** A **2** C **3** C **4** B **5** B **6** D **7** B **8** A **9** C **10** A

Page 18 **1** B **2** A **3** D **4** D **5** C **6** D **7** C **8** B **9** A **10** D

Page 19 **1** A **2** C **3** D **4** A **5** B **6** A **7** B **8** B **9** D **10** B **11** A **12** D **13** C

Page 20 **14 a** $4p + 1$ **b** $15 + 4d$ **c** $-3x$ **15 a** $8y$ **b** $30ab$ **c** $12a^2$ **16 a** $2a$ **b** b **c** $-p$ **17 a** $2y$ **b** 10 **c** $\frac{3p}{4}$ **18 a** $5y + 5$ **b** $3m + 3n$ **c** $6a - 10$ **19 a** $7p$ **b** $-2ab$ **20 a** $\frac{5t}{3}$ **b** $\frac{4a}{5}$ **c** $\frac{7}{x}$ **21 a** $3(y - 4)$ **b** $4(a + 5)$ **c** $2(3x - 4)$ **22** $3p + 17$

Page 21 **1** D **2** C **3** B **4** B **5** C **6** C **7** A **8** A **9** D **10** D **11** A **12** A **13** C

Page 22 **14 a** $2ab - 5a$ **b** $-7m - mn$ **15 a** $\frac{1}{2d}$ **b** $\frac{1}{6}$ **c** $\frac{4a^2}{25b^2}$ **16 a** $\frac{11t}{9}$ **b** $\frac{1}{2}$ **17 a** $-14c - 7d$ **b** $2b^3 - b^2$ **c** $-2a + 3b$ **18 a** $13a - 22$ **b** $8y + 9$ **19 a** $-2(a + 1)$ **b** $(2a + b)(p + 3)$ **20 a** $x^2 + 5x + 6$ **b** $a^2 - a - 12$ **c** $6b^2 + 7b - 5$ **21** $\frac{23p - 3}{6}$

Page 23 **1** A **2** B **3** B **4** D **5** C **6** C **7** A **8** D **9** C **10** B **11** D **12** D **13** A

Page 24 **14 a** $2a^2 - 3b$ **b** $6xy - 3xy^2 - 2x^2y$ **15 a** $10x - 4$ **b** $1 - 2m$ **c** $3 - 2ab - b + a$ **16 a** 2 **b** $\frac{11 - c}{4}$ **c** $\frac{3p + 6}{5}$ **17 a** $2a^2 + 2b^2$ **b** $-4xy$ **18 a** $3p^2 + 16pq + 5q^2$ **b** $6m^2 + 3n^2$ **19** $-3(x + 1)$

Indices

Page 25 **1** C **2** B **3** D **4** B **5** C **6** D **7** A **8** A **9** B **10** D

Page 26 **1** A **2** D **3** A **4** D **5** B **6** B **7** B **8** C **9** D **10** A

Page 27 **1** B **2** D **3** A **4** A **5** D **6** B **7** A **8** D **9** B **10** B **11** C **12** C **13** D

Page 28 **14 a** a^7 **b** $4y^3$ **c** $18p^7$ **15 a** x^3 **b** a **c** y^2 **16 a** m^6 **b** p^{18} **c** $4w^6$ **17 a** 1 **b** 2 **c** 1 **18 a** g^{12} **b** $24a^{11}$ **c** $12y^8$ **19 a** 4 **b** 6 **c** 4 **20 a** 2^4 **b** p^3q^2 **c** a^4b^5 **21 a** $-6a^6$ **b** $3m^3$ **c** p^4 **22 a** m^5n^3 **b** x^2y^3 **c** $9c^6d^4$

Page 29 1 C 2 B 3 C 4 D 5 C 6 B 7 A 8 C 9 C 10 C 11 D 12 D 13 A

Page 30 **14 a** $15m^6$ **b** $-18a^3b^7$ **c** $9x^{10}$ **15 a** $-3a^3$ **b** b^{-1} **c** $5p^2$ **16 a** $2m^8 + 4m^3$ **b** $a^2b^5 - 3ab^4$ **c** $x^7y^4 - 2x^9y$ **17 a** b **b** p^7 **c** s^{-6} **d** a **e** $m^{\frac{1}{5}}$ **f** p^2 **18 a** $\frac{1}{8}$ **b** $\frac{8}{5}$ **c** $\frac{9}{4}$ **d** 2 **e** 10 **f** $\frac{5}{2}$ **19 a** 6.3×10^{-4} **b** 8.9×10^{-2} **c** 9.7×10^{-4} **20 a** $x = 10$ **b** $x = 6$ **c** $x = 5$

Page 31 1 C 2 C 3 D 4 A 5 D 6 B 7 A 8 B 9 C 10 D 11 C 12 A 13 B

Page 32 **14 a** x^4y^6 **b** a^3b^9 **c** $\frac{1}{9p^4}$ **15 a** $\frac{3}{k^2}$ **b** $\frac{1}{2y}$ **c** $\frac{2}{\sqrt{m}}$ **16 a** 81 **b** 1 **c** $\frac{1}{2}$ **17 a** $x = 1$ **b** $a = -\frac{1}{2}$ **c** $y = \frac{3}{4}$ **18 a** 2^x **b** 2^3 **19** $\frac{y}{3x^2}$

Linear Equations

Page 33 1 B 2 C 3 B 4 C 5 A 6 A 7 C 8 B 9 D 10 D

Page 34 1 A 2 C 3 A 4 A 5 B 6 D 7 D 8 B 9 B 10 D

Page 35 1 C 2 B 3 D 4 B 5 C 6 C 7 A 8 D 9 D 10 C 11 A 12 A 13 C

Page 36 **14 a** $a = 4$ **b** $p = 5$ **c** $m = 1$ **15 a** $p = 13$ **b** $t = 17$ **c** $d = -11$ **16 a** $b = 7$ **b** $m = 2$ **c** $n = -2$ **17 a** $a = 7$ **b** $p = 3$ **c** $k = 1$ **18** $z = 8$

Page 37 1 D 2 A 3 B 4 A 5 A 6 B 7 B 8 B 9 D 10 B 11 D 12 B 13 C

Page 38 **14 a** $t = -3$ **b** $a = 2\frac{1}{2}$ **c** $m = 1$ **15 a** $q = -2$ **b** $p = \frac{1}{3}$ **c** $y = -1$ **16 a** $x = 5$ **b** $p = -14$ **c** $a = -3$ **17** $x = 1, y = 2$

Page 39 1 B 2 C 3 C 4 C 5 B 6 B 7 D 8 D 9 A 10 B 11 A 12 B 13 A

Page 40 **14 a** $a = 4$ **b** $k = -1$ **c** $p = -3$ **15 a** $y = -4$ **b** $b = 6$ **c** $a = 1$ **16 a** $x = -2, y = -3$ **b** $a = -2, b = 3$ **c** $m = -3, n = -1$ **17** 1450 adults, 550 children

Inequalities and Formulae

Page 41 1 B 2 C 3 D 4 A 5 B 6 A 7 B 8 C 9 D 10 C

Page 42 1 A 2 D 3 D 4 B 5 A 6 A 7 D 8 D 9 C 10 B

Page 43 1 D 2 B 3 B 4 A 5 A 6 B 7 D 8 A 9 D 10 B 11 D 12 A 13 A

Page 44 **14 a** **b** **c** **15 a** $x > 5$ **b** $x \leq 2$ **c** $x < -1$ **16 a** $m > 2$ **b** $p \geq 3$ **c** $a < 5$ **17 a** $y > -2$ **b** $a \leq -3$ **c** $q > -24$ **18 a** 36 **b** 45.38 **c** $7\frac{3}{5}$ **19** 86

Page 45 1 D 2 A 3 C 4 C 5 B 6 D 7 A 8 C 9 A 10 A 11 A 12 D 13 C

Page 46 **14 a** $a < -2$ **b** $c \geq -1$ **c** $p \leq -5$ **15 a** $b > -12$ **b** $m > 13$ **c** $y \geq 3$ **16 a** $p \leq 3.5$ **b** $a > 5$ **c** $q \leq 2$ **17 a** $t = 16$ **b** $a = 4$ **18** $y = \frac{2x - 13}{3}$

Page 47 1 B 2 C 3 B 4 C 5 C 6 A 7 B 8 A 9 A 10 C 11 D 12 B 13 D

Page 48 **14 a** $x < \frac{3}{4}$ **b** $a \leq \frac{5}{11}$ **c** $m < 1\frac{1}{7}$ **15 a** $y = \frac{2x + z}{\pi + 2}$ **b** $y = \frac{xz}{2x + z}$ **c** $y = \frac{xz}{x - z}$ **16 a** $s = 10$ **b** $h = 4$ **c** $c = 15$

Linear Relationships

Page 49 1 C 2 C 3 B 4 B 5 D 6 C 7 D 8 A 9 B 10 A

Page 50 1 C 2 D 3 D 4 D 5 A 6 A 7 C 8 B 9 B 10 C

Page 51 1 B 2 A 3 D 4 D 5 C 6 C 7 C 8 A 9 B 10 A 11 C 12 C 13 D

Page 52 **14 a**

x	0	1	2
y	3	4	5

b

x	0	1	2
y	4	3	2

c

x	0	1	2
y	1	3	5

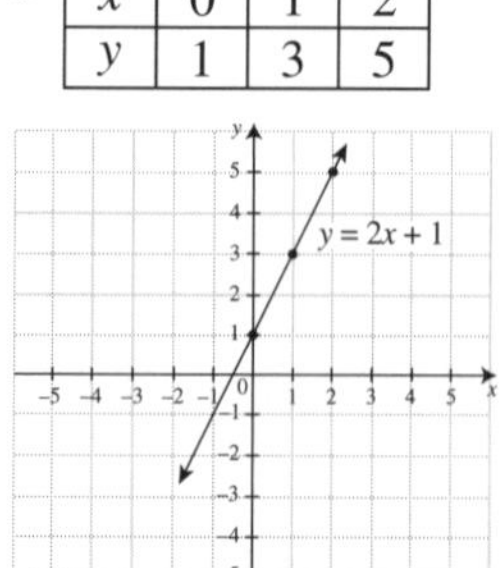

15

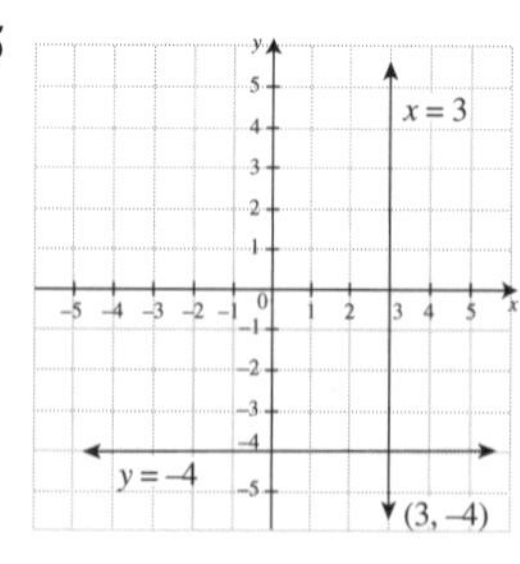

16 a 5 units **b** $\frac{3}{4}$ **c** $(2, 2\frac{1}{2})$ **17 a** $m = 3, b = 5$ **b** $m = -5, b = 2$ **c** $m = 1, b = 0$ **18 a** $y = 2x + 1$ **b** $y = 3x - 2$ **c** $y = -x$

Page 53 1 D 2 B 3 D 4 B 5 D 6 C 7 B 8 C 9 D 10 C 11 A 12 A 13 A

Page 54 14 a $\sqrt{65}$ units b $-\frac{4}{7}$ c $(-1\frac{1}{2}, -1)$ 15 a $m = 1, b = 3, y = x + 3$ b $m = -\frac{1}{2}, b = 2, y = -\frac{1}{2}x + 2$ c $m = \frac{3}{2}, b = -3, y = \frac{3}{2}x - 3$ 16 a

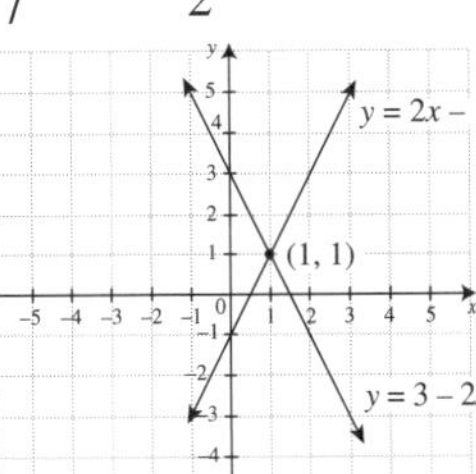

b

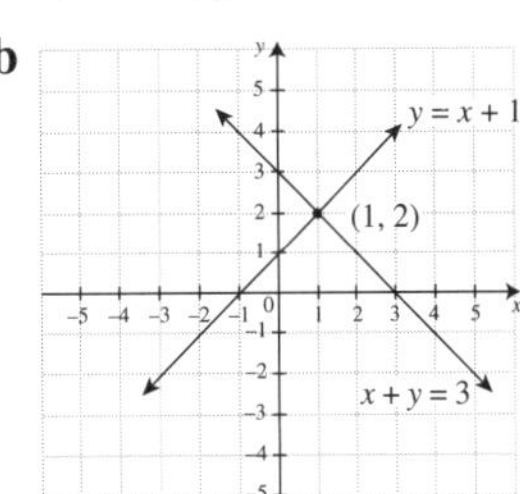

c

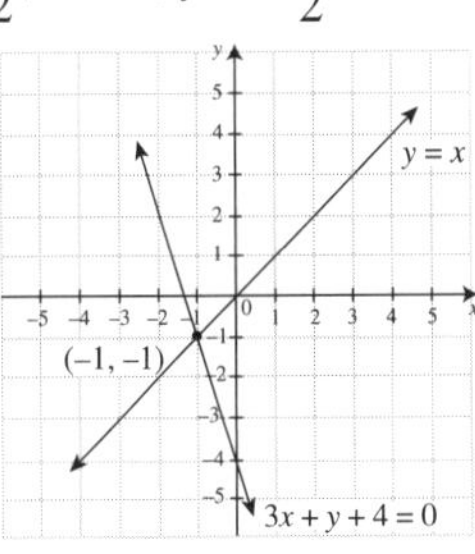

17 a $(-1, -9)$ b $(-1, 2)$

Page 55 1 B 2 C 3 C 4 C 5 D 6 D 7 A 8 C 9 D 10 A 11 D 12 A 13 A

Page 56 14 a

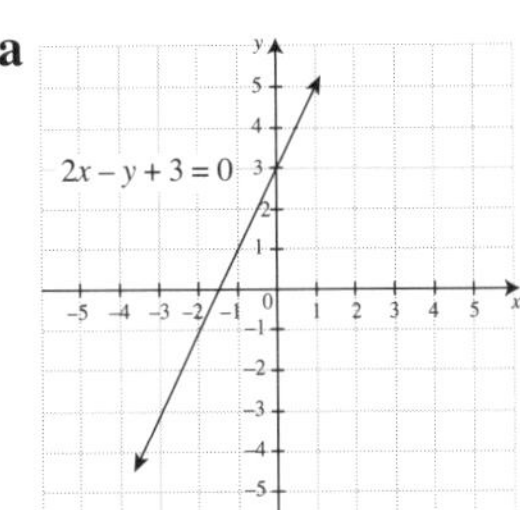

b

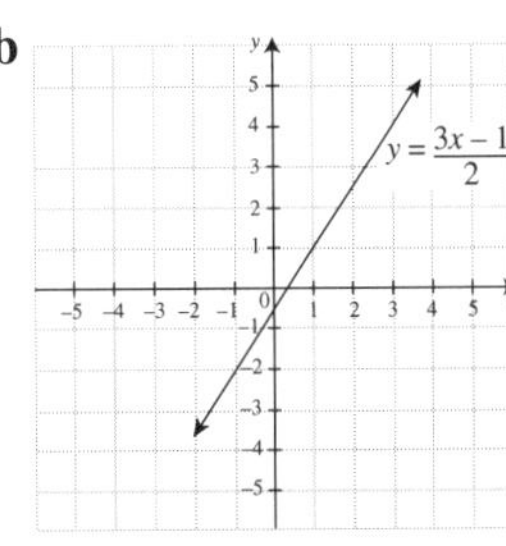

c

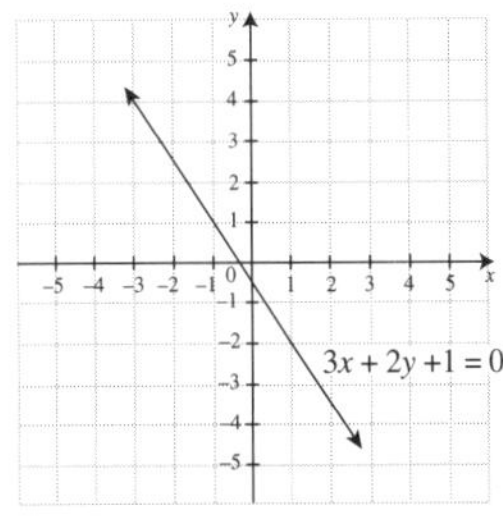

15 a $\sqrt{61}$ units b $(\frac{1}{2}, 2)$ c $-\frac{6}{5}$ 16 a $(1, 1)$ b $(2, 4)$ 17 a $y = -2x + 2$ b $y = -\frac{2}{3}x + 2$ c $y = 2x + 4$

Quadratics

Page 57 1 D 2 A 3 B 4 C 5 C 6 D 7 A 8 A 9 B 10 C

Page 58 1 B 2 C 3 C 4 A 5 D 6 A 7 D 8 B 9 D 10 B

Page 59 1 C 2 B 3 A 4 A 5 B 6 D 7 D 8 B 9 D 10 C 11 D 12 A 13 B

Page 60 14 a $x^2 + 8x + 15$ b $y^2 + 5y - 6$ c $a^2 - 3a + 2$ 15 a $2y^2 + 7y + 3$ b $5p^2 + 9p - 2$ c $6m^2 - 7m + 2$ 16 a $(a + 2)(a + 1)$ b $(w + 5)(w + 1)$ c $(b - 2)(b - 1)$ d $(x + 5)(x - 1)$ e $(y - 3)(y + 2)$ f $(z - 6)(z - 2)$ 17 a $x = -1, 2$ b $y = -4, 5$ c $m = 0, 2$ d $c = 3, -3$ e $a = 5, -5$ f $y = 0, 3$ 18 a $b = -3, -1$ b $n = 3, 4$ c $h = -5, 2$

Page 61 1 A 2 D 3 B 4 D 5 C 6 D 7 C 8 B 9 B 10 A 11 C 12 C 13 D

Page 62 14 a $6a^2 + 5a - 4$ b $8y^2 + 2y - 3$ c $12p^2 - 23p + 10$ 15 a $(p + 3)(p - 3q)$ b $(m - 5)(m - 3n)$ c $(y + 6z)(2y - 3)$ 16 a $2(b - 4)(b + 3)$ b $3(x - 5)(x - 3)$ c $5(q - 4)(q - 2)$ 17 a $n = 6, -1$ b $g = 9, 2$ c $p = 5, -2$ 18 a $\frac{x + 3}{2}$ b $\frac{4}{y - 2}$ c $m + 1$

Page 63 1 A 2 A 3 D 4 A 5 B 6 C 7 C 8 B 9 B 10 D 11 D 12 C 13 D

Page 64 14 a $4w^2 + 12w + 9$ b $9y^2 - 6y + 1$ c $25 - 4t^2$ 15 a $(3d + 4)(d - 2)$ b $(2x - 1)(x + 3)$ c $(3p - 4)(2p - 1)$ 16 a $m = -\frac{3}{2}, 4$ b $v = \frac{1}{4}, 5$ c $q = \frac{3}{5}, -\frac{3}{2}$ 17 a $\frac{a + 1}{a - 1}$ b 1 18 a $x = -3, y = -4$ or $x = 2, y = 6$ b $x = 4, y = 39$ or $x = -2, y = -3$

Rates and Proportion

Page 65 1 D 2 A 3 C 4 A 5 D 6 D 7 B 8 B 9 D 10 B

Page 66 1 B 2 A 3 D 4 D 5 B 6 D 7 D 8 C 9 B 10 A

Page 67 1 D 2 D 3 C 4 B 5 C 6 A 7 D 8 B 9 D 10 B 11 A 12 D 13 A

Page 68 14 a $y = 6x$ b 36 c 9 15 a 33.6 b 960 c 6 16 a 0.6 b 2 c 4.8 17 a $P = kn$ b $T = ks$ c $d = kA$ 18 a $C = kL$ b $L = kt$ c $C = ks^2$ 19 a $A = 8p$ b 160 m^2 c 15 L

Page 69 1 A 2 C 3 D 4 B 5 B 6 C 7 D 8 A 9 B 10 B 11 D 12 C 13 A

Page 70 14 a $k = \frac{1}{2}$ b $y = \frac{x}{2}$ c 4 d 12 15 a $d = 0.8s$ b 9.92 km c 7500 16 a $P = 12L^2$ b 27 c 6.4 17 a $m = 3.2r^3$ b 3.5 cm c 691.2 g 18 2.7 L

Page 71 1 B 2 A 3 D 4 D 5 D 6 A 7 B 8 A 9 C 10 C 11 C 12 A 13 D

Page 72 14 a multiplied by 4 b multiplied by 16 c divided by 4 15 a $E = \frac{400s^3}{27}$ b 691.2 kW c 4.2 m/s 16 a $k = 3.2$ b 2 c 8 17 a $r = \frac{7680}{l}$ b 128 c 32 cm 18 64

Number and Algebra Strand Tests

Page 73 1 C 2 A 3 D 4 C 5 D 6 A 7 A 8 C 9 B 10 D 11 D 12 B 13 D

Page 74 14 3.11 15 a 60% b 15% 16 \$77 866.80 17 \$1332 18 a \$1036 b \$6956 19 $10x - 3$ 20 a 1 b p^5q^4

21 a $x = 2$ b $q = 5$ 22 $x \leq -3$ 23 38 24 a 4 b −2 25

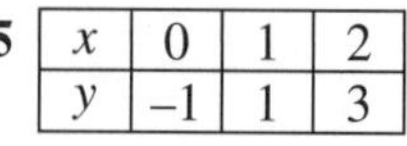

x	0	1	2
y	−1	1	3

26 a 5 units b $\frac{3}{4}$ c $(0, 3\frac{1}{2})$ 27 $2a^2 + 5a - 3$

28 $m = 5, -3$ 29 $C = kd$

Page 76 1 A 2 B 3 A 4 B 5 A 6 C 7 D 8 C 9 C 10 A 11 C 12 A 13 A

Page 77 14 a 5.9 b 0.24 c 0.066 15 \$4056.25 16 a \$87 160 b \$19 881.20 17 \$5680 18 $\frac{b^2}{6a}$

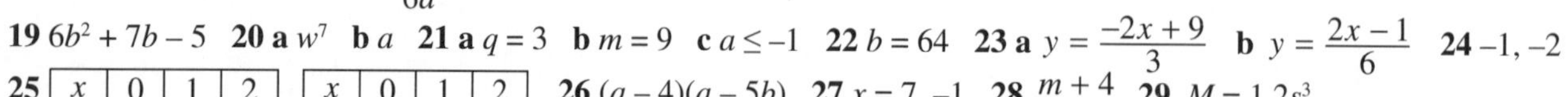

19 $6b^2 + 7b - 5$ 20 a w^7 b a 21 a $q = 3$ b $m = 9$ c $a \leq -1$ 22 $b = 64$ 23 a $y = \frac{-2x + 9}{3}$ b $y = \frac{2x - 1}{6}$ 24 −1, −2

25

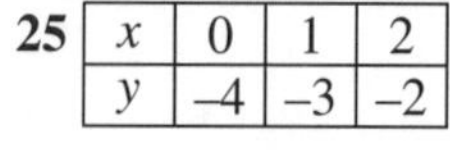

x	0	1	2
y	−4	−3	−2

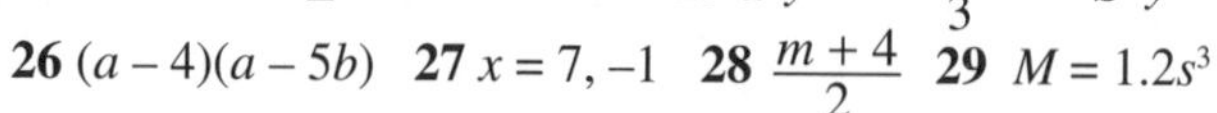

x	0	1	2
y	2	1	0

26 $(a - 4)(a - 5b)$ 27 $x = 7, -1$ 28 $\frac{m + 4}{2}$ 29 $M = 1.2s^3$

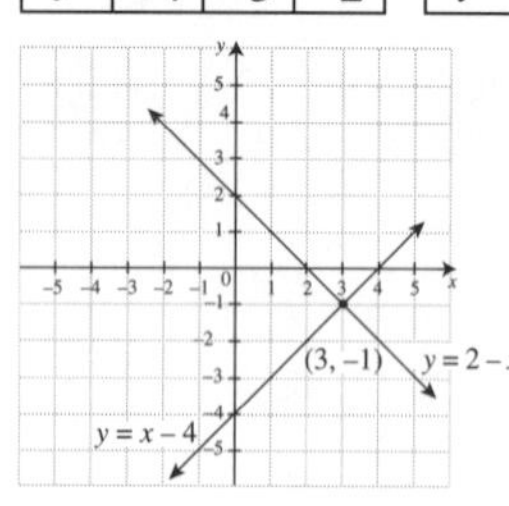

Page 79 1 A 2 D 3 D 4 A 5 A 6 D 7 A 8 A 9 B 10 C 11 C 12 A 13 D

Page 80 14 a 4.77×10^5 b 8.37×10^{-2} 15 $\frac{47}{90}$ 16 \$251.37 17 4% 18 $\frac{16c - 19}{6}$ 19 $-4pq$ 20 $8a^6b^9$ 21 $p = \frac{1}{4}$ 22 $x = 6, y = 1$ 23 $y \leq -\frac{2}{9}$ 24 a $\sqrt{80}$ units b −2 c (−1, 1) 25 $y = 22 - 5x$ 26 $\frac{m - 3}{3}$ 27 (1, 1), (−2, 19) 28 a $P = \frac{96}{\sqrt{x}}$ b $x = 16$

Pythagoras' Theorem

Page 82 1 A 2 B 3 D 4 B 5 D 6 C 7 C 8 C 9 B 10 D

Page 83 1 A 2 A 3 D 4 C 5 A 6 B 7 C 8 B 9 C 10 C

Page 84 1 A 2 B 3 C 4 A 5 C 6 C 7 B 8 D 9 D 10 C 11 B 12 D 13 A

Page 85 14 a 6.71 b 4.47 c 11.40 15 a 10 b 17 c 13 d 3 e 40 f 7 16 a 11.18 b 7.21 c 8.49 d 6.71 e 8.37 f 5.23

Page 86 1 D 2 B 3 D 4 A 5 D 6 A 7 B 8 D 9 C 10 C 11 C 12 D 13 B

Page 87 14 a $\sqrt{96}$ b $\sqrt{313}$ c $\sqrt{192}$ 15 a 5.3 b 23 c 21 16 a $a = 18.601, b = 13.601$ b $x = 4.472, y = 8.307$ c $p = 15.811, q$ 15.297 17 30 km

Page 88 1 B 2 A 3 B 4 B 5 B 6 A 7 C 8 A 9 C 10 D 11 A 12 D 13 A

Page 89 14 a 5 cm b 13 cm c 65 cm^2 15 a 7.81 cm b 15.62 cm^2 c 8.77 cm 16 a 3 cm b $\sqrt{116}$ cm c $\sqrt{125}$ cm 17 a 42 cm b 441π cm^2 18 a 4 units b 54 units^2

Area, Surface Area and Volume

Page 90 1 C 2 A 3 D 4 C 5 D 6 B 7 C 8 D 9 C 10 A

Page 91 1 B 2 A 3 B 4 A 5 B 6 D 7 B 8 D 9 B 10 C

Page 92 1 A 2 B 3 A 4 A 5 A 6 D 7 D 8 C 9 C 10 C 11 D 12 C 13 C

Page 93 14 a 125 cm^3 b 25 cm^2 c 150 cm^2 15 a 480 cm^3 b 48 cm^2 c 376 cm^2 16 a 960 cm^3 b 200 cm^2 c 672 cm^2 17 a 2.304 L b 1.656 L c 0.594 L

Page 94 1 D 2 D 3 B 4 C 5 B 6 A 7 A 8 C 9 D 10 A 11 B 12 C 13 C

Page 95 14 a 82.69 cm^2 b 92.73 cm^2 c 41.89 cm^2 15 a 808 cm^2 b 717.21 cm^2 c 494 cm^2 16 a 1440 cm^3 b 1272.35 cm^3 c 660 cm^3

Page 96 1 A 2 B 3 B 4 A 5 B 6 D 7 C 8 C 9 C 10 D 11 A 12 D 13 B

Page 97 14 a 15.52 cm^3 b 402.12 cm^3 c 100.69 cm^3 15 a 49.61 cm^2 b 330.57 cm^2 c 158.21 cm^2 16 a 256.19 cm^2 b 432 cm^2 c 211.21 cm^2

Trigonometry

PAGE 98 **1** B **2** C **3** A **4** B **5** B **6** A **7** A **8** C **9** C **10** A

PAGE 99 **1** B **2** D **3** D **4** B **5** C **6** D **7** C **8** A **9** D **10** A

PAGE 100 **1** C **2** B **3** D **4** A **5** B **6** C **7** D **8** A **9** B **10** A **11** C **12** B **13** D

PAGE 101 **14 a** 8.09 **b** 7.51 **c** 7.20 **d** 3.51 **e** 51.45 **f** 4.09 **15 a** 27° **b** 39° **c** 68° **16 a** 53° **b** 37° **c** 39° **d** 29° **e** 30° **f** 50°

PAGE 102 **1** D **2** A **3** D **4** C **5** B **6** D **7** C **8** B **9** A **10** C **11** A **12** D **13** A

PAGE 103 **14 a** 21.30 **b** 15.14 **c** 53°8′ **d** 52°36′ **e** 6.61 **f** 32.52 **15 a** 1.96 m **b** 216 cm **c** 12.38 cm **16 a** $\frac{3}{5}$ **b** $\frac{5}{12}$ **c** $\frac{3}{\sqrt{11}}$

PAGE 104 **1** D **2** B **3** C **4** A **5** D **6** D **7** C **8** B **9** A **10** A **11** B **12** C **13** C

PAGE 105 **14 a** $x = 36.91$, $y = 24.89$ **b** 67°59′ **c** 18.8 **15 a** 19.07 **b** 129° **c** 101° **16 a** 35° **b** 51° **c** 18°26′

Similarity

PAGE 106 **1** B **2** C **3** C **4** B **5** A **6** D **7** A **8** B **9** D **10** A

PAGE 107 **1** C **2** D **3** B **4** B **5** A **6** A **7** B **8** A **9** D **10** A

PAGE 108 **1** A **2** B **3** B **4** D **5** C **6** A **7** C **8** A **9** D **10** B **11** D **12** C **13** D

PAGE 109 **14 a i** 2 **ii** *SR* **iii** *ABC* **iv** *PQ* **v** 2 **vi** 8 **b i** 3 **ii** *FE* **iii** *YZW* **iv** *ZY* **v** 15 **vi** 4 **15 a** no **b** yes **c** yes **16 a i** 6 **ii** 4 **iii** 5 **b i** 5 **ii** 10 **iii** 21 **17 a** 12 m **b** 9.6 m **c** 4 m

PAGE 110 **1** B **2** A **3** D **4** A **5** A **6** A **7** B **8** A **9** D **10** C **11** B **12** A **13** C

PAGE 111 **14 a** *ABC* **b** *ZYT* **c** *LNM* **15 a** $a = 15$ **b** $b = 6\frac{2}{3}$ **c** $c = 4\frac{2}{3}$ **d** $d = 3\frac{1}{3}$ **e** $e = 6$ **f** $f = 1$ **16 a** $x = 10$ **b** 2 : 3 **c** 4 : 9 **17 a** $p = 4$ **b** 1 : 4 **c** 1 : 8

PAGE 112 **1** D **2** C **3** D **4** A **5** D **6** A **7** B **8** A **9** B **10** A **11** A **12** B **13** B

PAGE 113 **14 a** $x = 2.5$ **b** $x = 2\frac{2}{3}$ **c** $x = 15.6$ **15 a** $a = 4.8$, $b = 6.4$, $c = 3.6$ **b** $a = 15$, $b = 12$, $c = 20$ **c** $a = 2.4$, $b = 1.8$, $c = 3.2$ **16 a** 64.8 cm **b** 3.15 m **c** 3.6 m

Measurement and Geometry Strand Tests

PAGE 114 **1** D **2** B **3** C **4** D **5** B **6** D **7** D **8** A **9** A **10** C **11** B **12** A **13** D

PAGE 115 **14 a** 13.89 **b** 10.20 **c** 11.25 **15** yes **16 a** 168 cm^3 **b** 10 cm **c** 216 cm^2 **17** 572 cm^3 **18 a** 11.80 **b** 8.00 **19 a** 38° **b** 36° **20 a** 3 **b** *CD* **c** *SPQ* **d** *QR* **e** $m = 3$ **f** $n = 15$ **21** $a = 8$, $b = 4.5$ **22 a** 18 m **b** 10.8 m

PAGE 117 **1** B **2** B **3** C **4** C **5** B **6** C **7** B **8** A **9** A **10** D **11** C **12** C **13** A

PAGE 118 **14 a** 19.2 **b** 17.8 **15** 3.10 m **16 a** 2035.752 cm^3 **b** 994.858 cm^2 **17 a** $h = 3$ **b** 216 cm^2 **18** $\frac{4}{3}$ **19** 10.33 cm **20 a** *PQR* **b i** 8 **ii** 15 **c** 4 : 9 **21 a** $6\frac{6}{7}$ **b** $5\frac{1}{3}$ **22** $x = 16$ **23 a** 10 cm **b** 3 : 2 **c** 9 : 4

PAGE 120 **1** A **2** A **3** B **4** A **5** A **6** B **7** C **8** C **9** C **10** B **11** B **12** A **13** C

PAGE 121 **14** $a = 4$ **15** 20 km **16 a** $\sqrt{80}$ cm **b** $\sqrt{55}$ cm **c** 297 cm^3 **17 a** *A* has height of 2 m and *B* has height of $\sqrt{12}$ m. **b** *A* has volume of about 25.13 m^3 and *B* has volume of 27.71 m^3. **18** 18°51′ **19** $\frac{3\sqrt{3}a^2}{4}$ cm^2 **20** 13 km **21** 72 cm **22 a** $x = 6\frac{2}{3}$ **b** $y = 3\frac{3}{5}$ **23 a** The angles in ΔXYZ match the angles in ΔUWV (equiangular). **b** $p = 7.2$ **c** 43.2 cm^2

Statistics

PAGE 123 **1** B **2** C **3** B **4** B **5** A **6** B **7** A **8** B **9** C **10** A

PAGE 124 **1** A **2** B **3** C **4** B **5** D **6** C **7** C **8** A **9** C **10** D

PAGE 125 **1** A **2** C **3** D **4** C **5** C **6** C **7** A **8** D **9** B **10** C **11** B **12** B **13** C

PAGE 126 **14 a** 6 **b** 4 **c** 7 **d**

Scores	Freq. (f)	$f \times x$
5	1	5
6	6	36
7	3	21
8	4	32
9	2	18
Total	16	112

e 7 **15 a** 5 **b** 4 **c** 5 **d**

Scores	Freq. (f)	$f \times x$
2	5	10
3	1	3
4	3	12
5	7	35
6	3	18
Total	19	78

e 4.11 **16 a** 8 **b** 73 **c** 80 **d** 22 **e** 86 **f** 24 **17 a** 16 **b** 19.5 **c** 25 **d** 32 **e** 40

PAGE 127 **1** A **2** B **3** B **4** D **5** B **6** D **7** C **8** A **9** C **10** D **11** B **12** D **13** C

PAGE 128 **14 a** 4 **b** 17 **c** 17 **d** 15 **e** 17.5 **f** 2.5 **15 a** 4 **b** 14 **c** 14 **d** 12 **e** 14 **f** 2

16 a

Class	Class Centre	Freq.	fx
0–6	3	3	9
7–13	10	5	50
14–20	17	7	119
21–27	24	6	144
28–34	31	9	279
Totals		30	601

b 20.03 **c** 70%

17 a 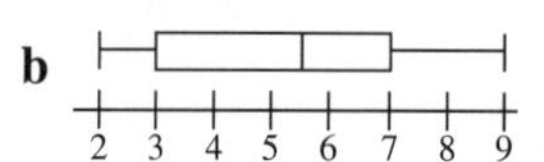

b 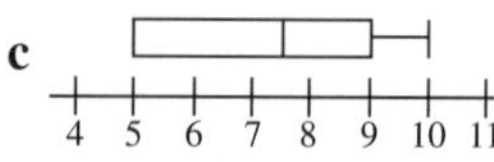

c
4 5 6 7 8 9 10 11

PAGE 129 **1** A **2** C **3** A **4** A **5** A **6** D **7** B **8** A **9** C **10** C **11** B **12** B **13** B

PAGE 130 **14 a** 40 **b i** remains at 15 **ii** decreased by about 3.6 **iii** decreased by 2 **15 a** Maths (5) is 1 less than Science (6). **b** Maths (17) is 6 more than Science (11). **c** Science is positively skewed; Maths is negatively skewed. **16 a** Boys (6) is 2 more than girls (4). **b** Boys (7) is 3 more than girls (4). **c** Boys is negatively skewed; girls is positively skewed. **17 a** $a = 4, b = 6, c = 8, d = 10$ **b** $p = 11, q = 13, r = 19, s = 21$

Probability

PAGE 131 **1** C **2** A **3** C **4** B **5** A **6** D **7** C **8** D **9** D **10** B

PAGE 132 **1** C **2** A **3** B **4** A **5** B **6** D **7** C **8** B **9** C **10** A

PAGE 133 **1** B **2** D **3** C **4** C **5** A **6** C **7** A **8** A **9** D **10** C **11** C **12** A **13** A

PAGE 134 **14 a** $\frac{1}{26}$ **b** $\frac{1}{13}$ **c** $\frac{1}{4}$ **d** $\frac{2}{13}$ **e** $\frac{3}{13}$ **f** $\frac{3}{4}$ **15 a** $\frac{1}{5}$ **b** $\frac{4}{5}$ **c** $\frac{7}{10}$ **16 a** 19

b

Total	Frequency	Relative frequency
2	3	0.03
3	5	0.05
4	8	0.08
5	11	0.11
6	14	0.14
7	a	0.19
8	13	0.13
9	10	0.1
10	7	0.07
11	6	0.06
12	4	0.04

c i 0.07 **ii** 0.27 **iii** 0.1 **iv** 0.4 **17 a** $\frac{12}{25}$ **b** $\frac{3}{25}$ **c** 0 **d** $\frac{22}{25}$ **18 a** $\frac{17}{40}$ **b** $\frac{2}{5}$ **c** $\frac{3}{10}$ **d** $\frac{7}{10}$ **19 a** $\frac{7}{40}$ **b** $\frac{2}{5}$ **c** $\frac{21}{40}$ **d** $\frac{1}{8}$

PAGE 135 **1** D **2** B **3** D **4** D **5** C **6** D **7** C **8** D **9** A **10** C **11** D **12** A **13** B

PAGE 136 **14 a** $\frac{1}{3}$ **b** $\frac{2}{3}$ **c** 1 **15 a** $\frac{1}{8}$ **b** $\frac{1}{4}$ **c** $\frac{7}{8}$ **d** $\frac{1}{8}$ **e** $\frac{1}{2}$ **f** $\frac{7}{8}$ **16 a** $\frac{1}{9}$ **b** $\frac{1}{9}$ **c** $\frac{2}{9}$ **d** $\frac{5}{9}$ **e** $\frac{4}{9}$ **f** $\frac{1}{9}$ **17 a** $\frac{1}{12}$ **b** $\frac{1}{36}$ **c** $\frac{1}{6}$ **d** $\frac{1}{6}$ **e** $\frac{1}{2}$ **f** $\frac{5}{6}$ **18 a** $\frac{1}{36}$ **b** $\frac{1}{4}$ **c** $\frac{1}{6}$ **d** $\frac{19}{36}$ **e** $\frac{5}{9}$ **f** $\frac{5}{9}$

PAGE 137 **1** C **2** A **3** A **4** B **5** B **6** D **7** C **8** A **9** C **10** D **11** B **12** C **13** B

PAGE 138 **14 a** 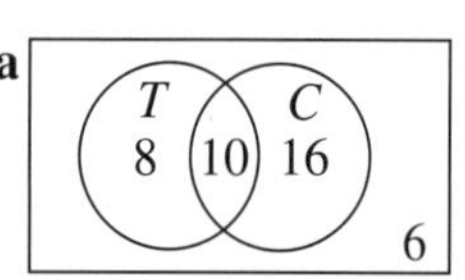

T: Tea C: Coffee

b $\frac{1}{4}$ **c** $\frac{7}{20}$ **15 a** 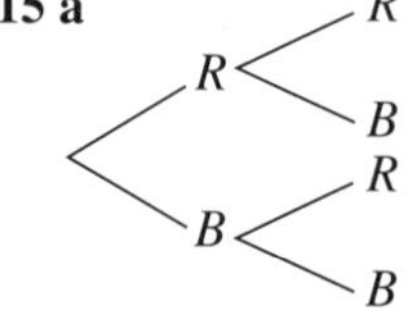

b $\frac{11}{18}$ **c** $\frac{1}{2}$ **16 a**

Driver's licence	Under 20	20 and over
yes	8	32
no	12	8

b $\frac{2}{15}$ **c** $\frac{1}{5}$ **17 a** $\frac{1}{6}$ **b** $\frac{1}{12}$ **c** $\frac{1}{36}$ **d** $\frac{2}{9}$ **e** $\frac{13}{18}$ **f** $\frac{5}{18}$ **18** $\frac{1}{4}$ **19** $\frac{5}{18}$ **20** 15

Statistics and Probability Strand Tests

PAGE 139 **1** C **2** D **3** C **4** B **5** B **6** C **7** D **8** B **9** A **10** D **11** D **12** A **13** B

PAGE 140 **14 a i** 26 **ii** 25 **b i** 19 **ii** 24 **c i** 22 **ii** 23 **15**

Class	Class Centre	Freq.
5–9	7	2
10–14	12	2
15–19	17	6
20–24	22	5
Total		15

16 a 7 **b** 5 **c** 4 **d** 4.5 **e** 3 **f** 5 **17 a** 12 **b** 3

18 a

Vowels	Frequency	Relative frequency
a	26	0.26
e	32	0.32
i	18	0.18
o	16	0.16
u	8	0.08
Total	100	1.00

b i 0.26 **ii** 0.58 **iii** 0.84 **19 a** 70 **b i** $\frac{9}{14}$ **ii** $\frac{23}{70}$ **iii** $\frac{4}{35}$ **iv** $\frac{5}{14}$ **20 a**

	Boys	Girls	Total
Overseas	8	12	20
Not overseas	6	4	10
Total	14	16	30

b i $\frac{4}{15}$ **ii** $\frac{2}{15}$ **iii** $\frac{2}{3}$ **iv** $\frac{8}{15}$ **21 a** $\frac{1}{4}$ **b** $\frac{1}{26}$ **c** $\frac{2}{13}$ **d** $\frac{9}{13}$

PAGE 142 **1** D **2** A **3** C **4** A **5** C **6** A **7** A **8** B **9** D **10** B **11** C **12** B **13** A

PAGE 143 **14 a** 12 **b** 12 **c** 12 **d** 13 **e** 1 **f** 12.47 **15 a**

Scores (x)	Freq. (f)	fx
10	4	40
11	2	22
12	4	48
13	0	0
14	6	84
Total	16	194

b i 14 **ii** 12.125 **iii** 12 **16 a**

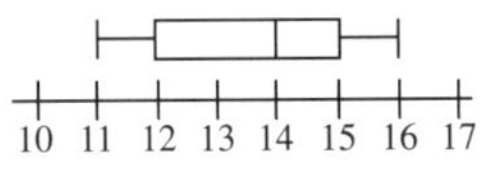

b

18 20 22 24 26 28 30 32

17 a

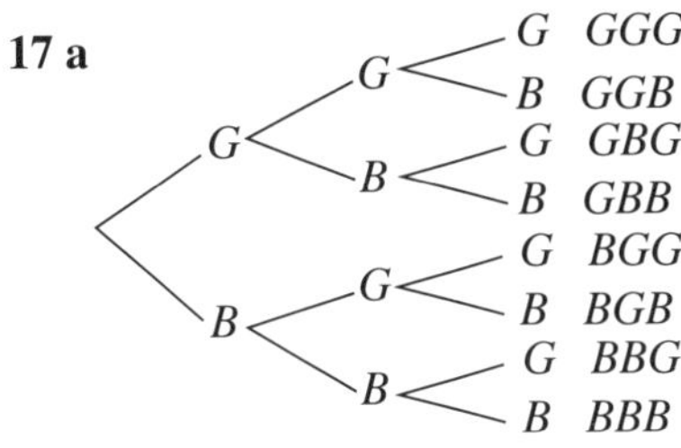

b i $\frac{1}{4}$ **ii** $\frac{3}{8}$ **iii** $\frac{1}{2}$ **iv** $\frac{1}{2}$ **18 a** 12 **b i** $\frac{1}{12}$ **ii** $\frac{1}{3}$ **iii** $\frac{1}{6}$ **19 a** $\frac{1}{4}$ **b** $\frac{1}{2}$ **20 a**

	1	2	3	4	5	6
1	1	2	3	4	5	6
2	2	2	3	4	5	6
3	3	3	3	4	5	6
4	4	4	4	4	5	6
5	5	5	5	5	5	6
6	6	6	6	6	6	6

b i $\frac{7}{12}$ **ii** $\frac{1}{4}$ **iii** $\frac{17}{36}$ **iv** $\frac{1}{2}$

PAGE 145 **1** D **2** C **3** C **4** A **5** B **6** D **7** A **8** A **9** B **10** A **11** B **12** C **13** D

PAGE 146 **14 a**

Home deliveries

Supermarket A		Supermarket B
7651	1	4
73	2	1367
81	3	5578
20	4	5

b i same range for both supermarkets (31) **ii** The median of B (31) is 6 higher than A (25). **iii** The IQR of B (14) is 8 lower than A (22). **15 a i** median: 14.0, $Q_1 = 13.4$, Q_3: 14.4, min.: 12.6, max.: 14.6 **ii** median: 12.8, $Q_1 = 12.6$, Q_3: 13.6, min.: 12.2, max.: 14.2 **b**

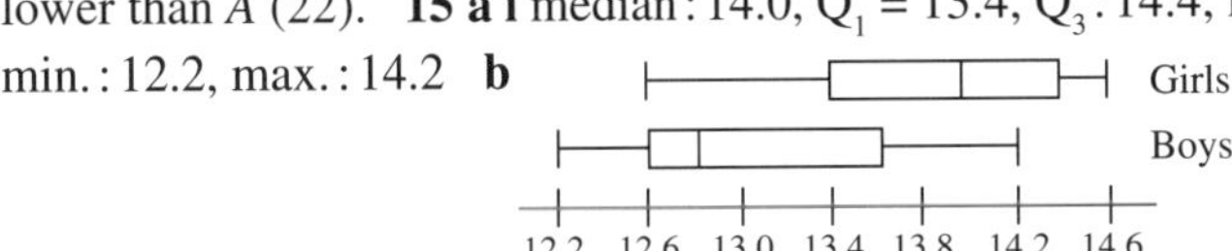

c i. The IQR (1.0) is the same for both. **ii.** The girls' data is negatively skewed, while the boys' data is positively skewed.

16 $a = 17, b = 24, c = 27, d = 28$ **17 a i** $\frac{1}{30}$ **ii** $\frac{1}{30}$ **iii** $\frac{1}{30}$ **b i** $\frac{117}{145}$ **ii** $\frac{27}{145}$ **18 a** $\frac{1}{18}$ **b** $\frac{7}{36}$ **19 a** 36 **b** 28

20 a $\frac{2}{9}$ **b** $\frac{1}{252}$

Exam Paper 1

PAGE 148 1 B 2 A 3 B 4 B 5 C 6 D 7 B 8 C 9 C 10 D 11 A 12 D 13 A 14 A 15 C 16 B 17 B 18 C 19 A 20 A 21 B 22 D 23 C 24 D 25 C

PAGE 150 **1** 5.2459×10^2 **2** 80% **3** \$18.90 **4** \$308.80 **5** \$80 810 **6** $3y - z$ **7** $10a^2 - 20a$ **8** $6x^4$ **9** $18a^5b^5$ **10** $a = -12$ **11** 1 **12** $x > -3$ **13** $y < 7$ **14** -2 **15** $g = \pm 4$ **16** $(x + 9)(x - 2)$ **17** $n = 5$ **18** \$44.72 **19** 11.18 **20** 600 cm² **21** 168 cm³ **22** $\frac{p}{q}$ **23** 8 cm **24** $\frac{1}{2}$ **25** NPQ **26** 8 **27** 95 **28** 17 **29** $\frac{3}{4}$ **30** $\frac{1}{2}$

PAGE 152 **1**

Fraction	Decimal	Percentage
$\frac{7}{25}$	0.28	28%
$\frac{1}{20}$	0.05	5%
$\frac{3}{40}$	0.075	$7\frac{1}{2}$%

2 \$1141 **3** $6a$ **4** $-p^2$ **5** 3 **6 a** $g = 2$ **b** $a = 3$ **7** 92 **8**

x	0	1	2
y	-2	1	4

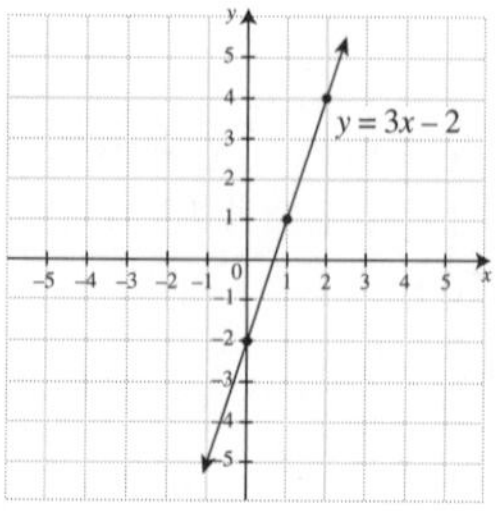

9 $2a^2 + a - 15$ **10** $p = 4, 2$ **11 a** $m = 6e$ **b** 30 **12** 6.708 **13** 126 cm³ **14** 5.18 **15** 40° **16** 2.5 m **17 a** 23 **b** 70 **18 a** 30 **b i** $\frac{3}{10}$ **ii** $\frac{1}{2}$

Exam Paper 2

PAGE 154 1 D 2 C 3 B 4 D 5 D 6 C 7 A 8 C 9 B 10 A 11 C 12 A 13 C 14 A 15 D 16 D 17 B 18 B 19 B 20 B 21 A 22 D 23 C 24 A 25 C

PAGE 156 **1** 1.380×10^7 **2** 61.7 kg **3** \$648 **4** $2c^2 - 5c - 3$ **5** $\frac{2}{p}$ **6** $12w^5 - 20w^3y$ **7** 1 **8** $\frac{1}{8}$ **9** $q = -\frac{1}{2}$ **10** $x = \pm 2$ **11** 11 **12** $x < 5$ **13** (1, 3) **14** -4 **15** $y = 4$ **16** $a = 7, -3$ **17** $4(x - 2)(x + 1)$ **18** $\sqrt{145}$ cm **19** 260.53 cm² **20** 628.32 cm³ **21** 3π cm² **22** 17.77 cm **23** $\frac{12}{13}$ **24** 9 **25** 14 **26** 12.5 **27**

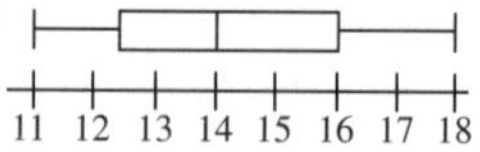

28 $\frac{6}{25}$

29

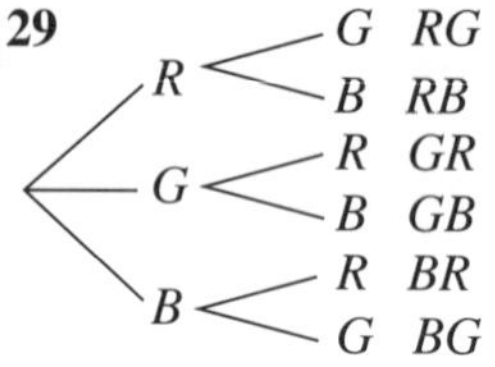

30 $\frac{2}{3}$

PAGE 158 **1** \$18 522 **2** \$483 **3** $5a^2 - 14a$ **4** $\frac{21a - 8}{6}$ **5** $2\frac{2}{3}$ **6 a** $m = -2$ **b** $w = 2$ **7** $a = 2, b = -1$ **8** $t = \frac{s - m}{u}$ **9** $\frac{a}{b + 2}$ **10 a** $\sqrt{80}$ units **b** $\frac{1}{2}$ **11 a** $P = kn^2, k = 3.6$ **b** 230.4 **12** $x = \sqrt{70}$ **13** 452 cm³ **14** 19 **15 a** 24 **b** 24 **16 a** $\frac{1}{6}$ **b** $\frac{1}{3}$ **c** $\frac{2}{3}$

Exam Paper 3

PAGE 160 1 C 2 B 3 B 4 D 5 D 6 B 7 C 8 D 9 C 10 B 11 A 12 C 13 A 14 B 15 A 16 A 17 A 18 C 19 A 20 C 21 D 22 D 23 D 24 D 25 A

PAGE 163 **1** $\frac{19}{20}$ **2** 0.658 **3** \$709.31 **4** \$75 106 **5** $\frac{x + y}{x}$ **6** $\frac{y^2}{3}$ **7** 3.17×10^{-5} **8** $2^a(2^b - 1)$ **9** $x = -1.8$ **10** $x = 1$ **11** $m = -3, n = -9$ **12** $A = \frac{C^2}{4\pi}$ **13** $x < -1$ **14** (6, –2) **15** $-4ab$ **16** $-2(x + 3)(x - 1)$ **17** 4 **18** 10 252 m **19** 12 cm **20** 572 cm² **21** 065° **22** 53° 8′ **23** 62.78 cm **24** CBD **25** $x = 6$ **26** $a = 4$ **27** 2.5° **28** $\frac{11}{12}$ **29** $\frac{11}{36}$ **30** $\frac{5}{14}$

PAGE 165 **1** $\frac{11}{15}$ **2** 7.4×10^7 **3 a** \$12 600 **b** \$17 136 **c** \$1844.50 **4** $24xy$ **5** $y = 2$ **6** $\frac{b^2}{2a}$ **7** $k = 5$ **8** $p = 3, q = -5$ **9** $y = 3x - 1$ **10** $m = -\frac{2}{3}, 1\frac{1}{2}$ **11 a** $d = \frac{120}{n}$ **b** 4 days **12** 13.4 cm **13** 733.49 cm³ **14 a** 8 cm² **b** 4 cm **15** 41°28′ **16 a** positively skewed **b** The Q_1, Q_3 and median would increase but the range would be unchanged. **17 a** $\frac{2}{5}$ **b** $\frac{1}{10}$